Science and Spirit

In relation and what that means

Michael J Spyker

AgapeDeum

Published in Adelaide, Australia by AgapeDeum
Contact: agapedeum.com

ISBN 978-0-6486957-3-8

Copyright © Michael J Spyker 2019

All right reserved. Other than for the purpose and subject to the conditions prescribed under the *Copyright Act*, no part of this publication may be reproduced, stored in a retrieval system, or transmitted in any form or by any means, electronic, mechanical, photocopying, recording and otherwise, without prior permission of the publisher.

This edition published in 2020

Publication assistance by Immortalise
Cover design: Ben Morton

CONTENT

1. The liberation of the Western mind — 1
2. A place for God — 16
3. Science and religion — 29
4. Science and personhood — 42
5. Spirit and personhood — 54
6. In a world of logic — 67
7. Space and time — 82
8. The nature of spirit — 94
9. A fearsome power — 106
10. A relational universe — 121
11. Science and spirit — 133
    References — 152

Chapter 1

The liberation of the Western mind

In our modern world there is a freedom of ideas that never existed to that extent before. Long established norms are progressively overthrown with always a next one to be dismantled. Sexual freedom is but one example. Relational validity, regardless of state or gender, another. The perception of nature has changed completely. Most notably that the universe was not created but somehow came into being of itself. Matter is now essentially a manifestation of energy. There is the suggestion of multiple universes. Philosopher Friedrich Nietzsche (1844-1900) decided years ago that God is dead. With so much evil in our world, many people readily accept that.

There are multiple reasons for these changes with rapid scientific achievement a foremost contributor. The corresponding technological developments create a very different world. Our globe has become a smaller place and readily accessible. People feel more self-sufficient. Less subject to the vagaries of life. They are better informed than their forebears, surely also because of the internet. Modern technology is incredible and rushing forward. Because science is conquering mysteries as large as our universe and as minute as quantum particles.

The modern mind is liberal and individualistic. Personal freedom is the norm. People think and communicate according to their convictions. There are many up-sides to this and also dangers. Worthwhile ideas can find ready expression, but so do the socially destructive ones. Our world is in flux. In the West morality has become a mostly private affair. The boundaries previously set by religion are falling over. Truth is in the mind of the beholder and no longer subject to Christian tradition.

Historically, the dictates of the church have been used to further ecclesiastical power. It happens still today. Modern people though walk more easily away from such influences. It is a fairly recent phenomenon. During the first three quarters of its existence the Church firmly ruled society. (The early centuries being an exception.) Dogma was rigidly enforced in the name of God. Wars were fought in that name. The Crusades and Inquisition are well-known examples. Belief itself was laden with the possibilities of condemnation. Inspirational thinking was restricted and the potential of philosophy and science remained hamstrung.

Knowledge was deemed acceptable when it aligned with Church teaching. It made philosophy subject to theology and science to how the Church viewed reality. Art depicted biblical scenes exclusively and nature as God had created it. Abstract ideas were of the devil. Ecclesiastical authority was the most powerful force in society and the Pope its prime authority. Many of those influences would these days be denounced as spiritual abuse.

The extremes were most apparent in medieval times, when the spiritual realm permeated every person's thinking. The sinful life and potential of hell was a serious burden to carry. This predicament could be solved through the Church, of course, with baptism the first step on the way of salvation. The Church preached a stark reality. For instance, unbaptised babies were considered to be eternally lost. So were folk who had fallen into serious sin. Much of what was considered serious, would be of little importance today. To a medieval person, the pitfalls of sin were everywhere. Walking through the dark at night was always dangerous. The devils would be lurking to take possession of the most dynamic aspect of personal being: the human will. Once that had become invaded, all was lost. How this state of affairs affected people is hard to imagine in modern times. The spiritual was a realm of angels and demons. They *really* filled the air. This belief found it fullest expression during the Gothic period (12th-16th century). It is well described in an observation about Gothic art by Oswald Spengler.

> The saints and angels float in the aether, but the devils leap and crouch and the witches rustle through the night. It is the contrast, day and night, which gives Gothic art its indescribable appeal – this was no artistic fantasy. Every man knew the world to be peopled with angels and devil troops.[1]

It was the scientists, who began to liberate the world from this oppression. It could cost them their lives. In 1600, Giordano Bruno was burned at the stake in Rome for his scientific ideas and unorthodox theological views. The great astronomer Galileo, tried for heresy in 1633, had to recant his proposals and spent his last nine years under house arrest. None of the early scientists denied a belief in God. But they championed a greater freedom of mind and challenged orthodoxy. Slowly but surely, over time, the authority of the Church was confronted with views that it could no longer erase. At first by the philosophers and scientists. They still confessed to belief in God. But possibly of an unorthodox kind.

Isaac Newton (1642-1727) is an example. Folklore tells us that he saw the apple fall from the tree and wondered why? It opened the door to scientific enquiry. Experiment was combined with massive strides forward in mathematics. Newton invented the calculus, though German philosopher and mathematician Gottfried Leibniz (1646-1716), working independently, claimed to have been first. Unbeknown to many, Newton was an accomplished theologian also. Well versed in biblical languages and Early Church history. He had a committed belief in God, but not in accordance to what the Church was teaching. Newton held to the heresy of Arianism. The conviction that the Son of God had not always existed, but at one point was begotten by the Father. The Godhead was not a Trinity as confessed in the Creed. Richard S. Westfall reports that,

> In Newton's eyes, worshipping Christ as God was idolatry …. Well before 1675, Newton had become an Arian in the original sense of the term. He recognised Christ as a divine mediator between God and humankind, who was subordinate to the Father who created him.[2]

Newton kept these view strictly private and held important positions in the Church. On his deathbed he refused the sacrament. As a scientists Newton was a creative and exceptional thinker coming to his own conclusions. With considerable theological competence, he applied that approach to biblical understanding. It moved him to disagree with Church teaching.

Isaac Newton, and later Charles Darwin's (1809-1882) evolution theory, were fundamental to the sciences and their development. Science began to untie the bonds placed upon it by theology. Church dogma became considered as questionable. There was a different way in which to explain reality. Not one based on a belief in what could not possibly be seen, but one anchored in facts and a reasoned interpretation. The Church began to lose its grip on society. It was facing a lasting problem.

Following science, philosophy undid the shackles of theology and became liberated into its own capacities. As a consequence people's self-perception changed. This shift in identity began in France in the late 17th century. The Enlightenment lasted for about one hundred years. People began to consider reality, including their own person, in ways other than religion. The driving force behind those cultural changes were writers, scientists and philosophers. At the forefront of that time was Rene Descartes (1596-1650). He coined the well-known phrase *ergo cogito sum* – 'I think therefore I am.' Descartes was a mathematician, who decided to make logic central to an explanation of life. In the tradition of Thomas Aquinas (1225-1274) and natural theology, he sought to prove the existence of God by reason. In true Enlightenment fashion, his focus was not on the handiwork of God, but on his personal psychology. He reflected on his own cognitive processes. These days Descartes' philosophical approach is called rational empiricism. He was a free thinking scientists who kept the Christian faith. Descartes had some bitter arguments with Dutch theologians and withheld from publishing some ideas so not to offend the Jesuit mentors of his youth.

In society a diminished adherence to Christian belief became unavoidable. Philosopher and eloquent writer Voltaire (1694-1778) felt free to often ridiculed Roman Catholicism. Over time many philosophers followed suit in declaring religion of no interest. It is the view of philosophy today.

A place for the spiritual is now mostly discarded in the disciplines of science, philosophy and psychology. Over a period of a few hundred years, and increasingly in the modern era, the relation between knowledge and religion has been reversed. From a mind restricted by the dictates of Christianity, the modern mind is one in which religious faith is considered irrelevant to the abilities of the intellect. And not just the intellect. Faith is not needed in living life altogether.

The sciences have engendered the idea that unscientific knowledge is not what *actual* knowledge is about. Unless insights are obtained using scientific methodology, it should not be considered reliable. This perspective holds non-scientific disciplines to be of lesser importance. In response, theology and philosophy have made an effort in showing that their disciplines do not lack rigorous method and thus qualify as scientific. It would ensure legitimacy in a mechanistic world. In this, philosophy has been more successful than theology. Karl Popper's falsifiability theory plus the paradigm shift of Thomas Kuhn, and Paul Feyerabend's focus on method, are three examples how philosophy might influence the sciences. Theology cannot claim this kind of interaction.

In his book *Theology and the Philosophy of Science* (1976) Wolfhart Pannenburg made an attempt. The title shows that theology should be seen as philosophically relevant to scientific endeavour. It applies distinct methods and consequently science should not disregard theological insights. They have a significant history and stood the test of time. Needless to say, science refuses to see the need of this. The idea of finding connectivity between the spiritual and scientific seems fanciful. The esoteric and material are poles apart. That problem will be addressed in *Science and Spirit*.

That the scientific fraternity is open to considering the value of philosophy is made possible because modern philosophers have limited the focus of their enquiry. They have replaced the majestic question of what it might mean to exist, with those attending to logic, language, ethics, politics and insights with a practical component.

Much to the dismay of philosophy enthusiast Brian Magee, as shown in his readable book, *Confessions of a Philosopher* (1997). By turning away from asking, 'What is truth?' and 'What is free will', philosophy aligns better with what the sciences consider relevant. After millennia of philosophical endeavour, no final answers has been given to key questions about human existence anyway. Modern philosophy has given up on it – it has become more 'useful.'

A good example of philosophy addressing science is Karl Popper's falsifiability theory. It addresses the serious problem of ensuring that a scientific theory is accurate. The most popular explanation of Popper's idea, which is quite extensive, concerns white swans. For years it was held in England that all swans are white, for that is what English swans are like. Until in Australia black swans were seen. So, the idea that swans were always white had been wrong – did not hold up as dependable theory. Consequently, Popper insisted that for a theory to be seen as correct, you had to imagine why it might not be. How can it be proven that swans are not always white? Simple: go find one of a different colour. If however hard you try, you cannot do so, then it may be assumed that your white swan theory is accurate. The theory is unfalsifiable. But you can never be absolutely sure. Future discoveries may still find swans that are not white. But at least you have given it your best. You have investigated a black swan possibility. Unless the question is asked, 'what might prove the theory wrong?' research will not qualify as comprehensive. Formulating that question correctly can be difficult though. Popper was struck by this idea while attending a lecture by Einstein who asked, 'Under what conditions would I admit that my theory is untenable?'[3]

While science may find that philosophy is of some use but not theology, Christian scientists would prefer to think otherwise. But finding theological relevance in their expertise is difficult. Even when the presumed scientific superiority over other ways of gaining knowledge is illusionary. Science will not unlock all the mysteries of the universe. As many scientists recognise. Richard Dawkins, who remains a forceful prophet of scientific superiority, annoys many colleagues. They don't ascribe to an all-conquering future knowing science's inability to penetrate the absolute essence of things. Quantum mechanics, the frontier of modern physics, is to be admired. But as an interpretation of the *very depth* of reality it comes up short.

Mathematics, the language of quantum, often involves approximations. For instance the value of pi is not exact for it represents 3 with an infinite row of numbers after the decibel point. So, the true circumference of a circle cannot ever be mathematically known. Many values in physics involve a statistical approximation of what is *really* out there. Einstein's concept of spacetime is an example. It is a mathematical model and not physically real. Mathematics is not itself a closed system either. Kurt Gödel's incompleteness theorems show the inherent limitations of arithmetic. Mathematical calculation cannot ever be totally conclusive. It does not detract from the enormous effectiveness of physics or its achievements. But absolute reality remains elusive.

Both science and philosophy involve logic. Research may include inspiration and intuition, but only as a stepping stone to the real work: that of calculating and testing. But there is more to reality and life overall than these activities. Comprehensive understanding of nature requires a broader perspective. The world is not explained by logical thinking alone.

Dutch philosopher Herman Dooyeweerd addresses this in his critique of theoretical thought. That logical thought can be autonomous is a fallacy, Dooyeweerd explains. His comments aimed at philosophy likewise apply to the sciences. Dooyeweerd insists that the mode of thinking involves the whole-of- person experience from which it cannot escape.

> How is the theoretical attitude of thought characterized? What is its inner structure by which it differs from the non-theoretical attitude of thinking? It displays an antithetic structure wherein the logical aspect of our thought is opposed to the non-logical aspect of our temporal experience.[4]

When a person thinks, it comprises of many fundamental modes. Dooyeweerd mentions at least a dozen of these. Including: a spacial aspect, a feeling and a cultural one, and also the aspect of faith.[5] In simple terms: thinking cannot be but subjective. It is never purely logical. As a Christian Dooyeweerd insists there is value in insights gained from revelation. Necessary value at that.

Logic has diminished that value in the modern world. Early 20th century quantum became the revolutionary new physics. It would conquer scientific understanding. Many of its theories built on discoveries that had been tested and found accurate. It developed those further mathematically. Often quantum theories cannot be proven empirically, for the reality addressed is subatomic and beyond verification. Yet many scientist are convinced of the validity of those mathematical theories. Logic is a powerful tool.

Among the leading proponents of quantum mechanics early on were Neils Bohr and Werner Heisenberg. They debated it with Einstein who remained sceptical. Erwin Schrödinger was another scientist of note at that time. His grandson Terry Rudolph, a quantum physicist who took his degree studies in Australia, writes that 'Maybe quantum mechanics doesn't capture everything that's really going on ….. the description of reality is ambiguous.'[6] His grandfather raised that problem years ago.

Quantum theory is where theology might best interact with physics. Science is aware that there is a deep reality quantum has been unable to penetrate. There is no reason to believe that it ever will for mathematics has its limits. This deepest reality is bound to be the presence of God by which all exists. But how to explain this connection into physics. The thinking required must make use of both theology and philosophy.

Thought experiments are common in science. Have an idea, think it through, and see where it may lead. That thinking is directed by the laws that apply to the experiment. For a thought experiment in theology that would be Scripture.

In part, *Science and Spirit* is such an experiment. It suggests the concept of spirit to be the foundational influence in creation. The connectivity between science and religion will always remain limited for their spheres of knowledge are different. But in considering the idea of spirit, and what that could mean with regard to the physical, there are possibilities. Recognising God's Spirit as an agency in our universe is not fanciful but scriptural.

The scientific detail in *Science and Spirit* will help those unfamiliar with it. The information is readily found in popular publications. For the Christian faith to increase its relevancy an open mind towards the sciences is essential. An honest appraisal of spirit and spirituality and their value to the modern person is of equal importance. Both topics are discussed in *Science and Spirit*.

Chapter 2

A place for God

All of reality is a matter of perspective. Isaac Newton saw an apple fall to Earth. He noticed movement and as a scientist interpreted it mathematically. Science has its ways in understanding a falling apple. A quantum physicist would perceive the fruit as a bundle of energy made up of myriads of particles that somehow manifest as an apple. Personally, I would consider the apple to be an appealing object that fits in the hand nicely. From a global perspective, the apple is just a tiny spot on an Earth that moves around the sun at great distance and with high speed. Apply a universe point of view and the apple basically disappears. How God sees the apple is beyond comprehension. All depends on perspective.

On a clear night in the open country, away from the haze that city light projects into the sky, I can get an idea of the size of our universe. Only a little, for the whole universe is exceptionally large. 'If the universe were shrunk so that Earth were the size of a period on this page, the centre of the Milky Way would be 5 million miles away.'[7] The actual distance, with the earth at full size, is mind blowing. This example concerns the earth and its galaxy the Milky Way only. It is suggested that there might be as many as two trillion galaxies. This guess is unverifiable. The existence of a very large number of galaxies though is a certainty.

The extensively large defies imagination. Likewise with quantum and the minutely small. A nanometre (nm) is a billionth of a metre. 'DNA, that tightly wound double helix of molecular instructions, is only about 2 nm wide – though if you unfurled the coils, the strands of DNA in a single cell would stretch over 2 m long.'[8] All that from a single cell. The cell consists of molecules and smaller again, atoms. Those 'atoms are generally a few tenths of a nanometre across, around 100-500 picometres.'[9] There are particles much smaller than an atom, like the Higgs boson, which is currently the smallest known particle. The very large and minutely small are well beyond any kind of reality our mind can make sense of.

It is no different with the spiritual. The revelations of Scripture can be staggering. Somehow, the very large and the tiny minute, hold together in Christ, who is God.[10] It is impossible to imagine the reality of that. Somehow though there must be an agency to make it happen. Ideally, an agency I can make some sense of, however limited. Jesus told the woman of Samaria that God is spirit.[11] I hold, that it is by spirit that our universe is created and exists. From God's Holy Spirit, creation comes forth by means of *universal spirit*. This universal spirit is energy-like and can manifest as substance. Science agrees that fundamentally all existence is an energy dynamic. Consequently, the idea that spirit can translate into matter is philosophically plausible. There is a place for God within scientific enterprise. Not one that the sciences are willing to consider though. They are busy in dealing with the 'real' problems of force fields and matter.

Scientific successes are plentiful but leave many questions unanswered. No scientist will deny this. It is unresolved questions that move the sciences onward and has done throughout history. Different ideas finding prominence in various ages. For millennia physics was determined by Aristotle's philosophy of space, time, motion, and cosmology. Until in the early sixteen hundreds when the Copernican revolution liberated physics from that ancient Greek take on reality. 'The culmination of that revolution was Isaac Newton's proposal of a new theory of physics, published in his *Philosophiae Naturalis Principia Mathematica* in 1687.'[12] But Newton's was an approach to physics that opened a gateway to yet another.

When Albert Einstein entered some two hundred years later, a next revolution occured. Its fruits continue into this day. It began in 1900 with Max Planck's demonstration that energy is not continuous but quantised. Planck assumed that his discovery could be reconciled with Newtonian physics. Einstein thought otherwise. His theory of relativity was a break-away from Newton and it made Einstein the father of modern physics.

Einstein calculated the speed of light at almost 300.000 km per second and the declared that nothing in our universe can move faster. Light is a phenomenon with an interesting history. Newton considered it a particle. His Dutch contemporary, Christian Huygens, disagreed and declared it a wave. The wave theory was considered correct for hundreds of years. Until in 1905 Einstein suggested that light was particle-like, cut up in little bits called light-quanta. The idea found little acceptance in the scientific community. In 1916 Einstein added the thought that light-quanta had momentum, which is a wave-like property. Within ten years Einstein's light-quantum was known as the photon.[13]

The answer as to whether light is a particle or a wave is confusing like so many phenomena in modern physics. It depends on the kind of measurement you take. The measurement determines the outcome. So, wave or particle? In reality, light is both. That is how science understands it. In its absolute reality, light might well be neither.

Einstein's influence on physics was enormous. At the end of his career it began to wane and his ideas came to be seen as not quite what progress was about. Physics had moved on in the direction of quantum mechanics, the realm of the very small. Quantum became a discipline with its own peculiar nature. Einstein could not agree with this development. His objections and discussions with Neils Bohr about a physics that focused solely on the atom and smaller, are well recorded. One theory that irked Einstein in particular was the uncertainty principle discovered by Werner Heisenberg. Lee Smolin explains.

> The *uncertainty principle* tells us that we cannot measure a particle's position and momentum at the same time. The theory yields only probabilities. A particle – an atomic electron say – can be anywhere, until we measure it; our observation in some sense determines its state. All of this suggests that quantum theory does not tell the whole story. As a result, in spite of its success, there are many

experts who are convinced that quantum theory hides something essential about nature that we need to know.[14]

Einstein subscribed to that notion. He famously declared that, 'God does not play dice with the universe.' He is known to have said: 'I want to know how God created this world. I am not interested in this or that phenomenon, in the spectrum of this or that element. I want to know His thoughts, the rest are details.'[15]

At the end of his life Einstein switched his attention to the unification of gravity and electromagnetics. This remains one of physics' major unresolved challenges. A single theory that incorporates both gravity and electromagnetics remains elusive. Quantum mechanics is no help in solving the problem. In physics there are now a number of major theories that explain reality in their own way. It means that none can be considered a full description of true reality, for that presents as a unified whole.

Physics proceeds in the hope – or some would say the certainty – that eventually a unifying theory will be found. Physics never lacks imagination nor shirks from being controversial. Predicted outcomes regularly become verifiable because of increased technological ability. But many theories will remain unproven for they are beyond of what anyone can ever test.

Einstein's theory of general relativity inspired the idea of the Big Bang. That our universe is the result of an initial explosion of matter that was condensed to the size of a small pea. The speed and force of that explosion are unimaginable. Somehow, that matter expanded into what physics is presently investigating: the incredible distances into the unknown at one end of the scale, and the unfathomably tiny intricacies of quantum at the other. Apparently this magnitude has come about naturally over time – a very long time. Big Bang theory is common knowledge. Many findings of cosmology have been proven dependable, but much remains unknown.

That our universe holds together at all is due to the proven consistency of the laws of physics across the cosmos. Science recognises it, but has no idea why it is so. Where do these laws come from? And, as has come to light recently, how consistent are they really? At the extreme ends of physics those laws begin to break down.

In 1995, investigating the nanokelvin zone, which is very close to absolute zero, physicists Carl Wieman and Eric Cornell noticed that rubidium atoms at those extremes shifted phase – not to a liquid, or a solid, but to an entirely new state of matter, never seen before.[16] Already in 1924, Einstein and Indian physicist Satyendra Nath Bose theorised that as individual atoms neared absolute zero, they would change in an extraordinary way.[17] These predictions confirm the consistency in the laws of physics for how otherwise could Einstein foresee this breakdown. But it remains confusing.

Visionary scientist Holger Bech Nielsen, the inventor of string theory, which these days is the most prominent theory in quantum physics, became an advocate for what he called *random dynamics*. Again, Smolin explains:

> Everything we think of as intrinsically true, such as relativity and the principles of quantum mechanics, he thinks are just accidental facts that are emergent from a fundamental theory so beyond our imagination that we might as well assume that its laws are random.[18]

Unsurprisingly, the idea of the present laws of physics being based on randomness found little acceptance. But it shows the possibility of a deeper reality into which science is unable to penetrate. From a Christian perspective, that would leave a place for the influences of God.

Nobody as yet has an inkling where life comes from. Nor, why molecules might stick together to make a stone. There are many questions that science has no answer to. To questions of a spiritual nature, such as the occurrence of life, love and morality, science expects to find an answer to through improved physics. The new frontiers of biology and chemistry, which heavily depend on quantum, are to offer satisfactory explanations one day. The sciences, in spite of their tendencies to be overly confident, are to be admired. They shun the spiritual for in that realm mathematics doesn't apply. Seemingly, science and spirit are incompatible. Insights about a worldview that allows for a measure of integration will need to come from religion.

This situation needs adequate attention from Christian theology. The science and religion debate must be flexible in finding concepts that leave a place for God within scientific understanding. Even if it only helps the Christian and leaves the sciences unperturbed. There is sufficient 'space' within the physical processes for a theological response.

Copernicus released physics from the ideas of

Aristotle. Theology might likewise liberate its discipline from these influences. Herman Dooyeweerd has addressed the issue showing that the duality of form and matter, fundamental to the worldview of ancient Greece, is unbiblical. The form and matter concepts eventuated over time in Greek consciousness due to two different approaches to religion. The religious practices of the lower classes, who mostly worked the land, focused spirituality on the sensual and nature. While the religious ideas of the cities regarded autonomous reason most highly. It developed a cult aimed at perfection of the mind. Eventually, these two separate ideas became united in philosophy. As matter (the land) and form (the mind), which resulted in an inescapable dualism.

Dooyeweerd observes about Greek philosophy:

> Since a real synthesis between the opposite motives of form and matter was not possible, there remained no other recourse than that of attributing the religious primacy to one of them with the result that the other was depreciated.[19]

Come Aristotle, god is conceived as pure form with the human mind being able to apprehend that reality. Pure form was eternal perfection unaffected by mere matter. It was the matter principle that became reduced to one of imperfection and decay; that of nature and the human body. There was a distinct separation between form and matter. The incorporation of these ideas into theology, Dooyeweerd considered unbiblical. Scripture has no special regard for the power of reason above other human abilities and sees it simply as one of many modal expression a person is capable of. Biblically, the seat of human reality is found in a person's heart. It encompasses all a person is about.

Thomas Aquinas (1225-1274) has been instrumental in applying Aristotelian philosophical concepts to the interpretation of Scripture. One concept that has done significant harm to the understanding of God is the Unmoved Mover. It encapsulates Greek dualism and suggests a God remote from creation. It is based on the notion that in creation movement is central to existence. If something moves, it belongs to creation. Consequently, wherever creation originates from must be unmoving otherwise it would be part of creation. That cannot be, for the divine is uncreated. Thus god is unmoving.

Karen Armstrong considers it one of Aristotle's less inspired achievements.[20] René Descartes is bold enough to suggest about Greek philosophers, that many of their ideas can hardly have been taken seriously by themselves.[21] The accuracy of that remains to be seen. But when dualism, this disconnected and static image of divinity, becomes integrated into the Christian perception of God, it does no end of damage to an affective – and effective - understanding of God. God as Unmoved Mover, fully distant from Creation, is a disturbing concept.

Spiritual theology, far less directed by the power of reason than dogmatics, has not taken the Aristotelian idea seriously. Meister Eckhart (1260-1329), for one, makes God's personal involvement with Creation absolutely clear and writes the following about suffering.

> However great your suffering may be, if it first passes through God, then he must first endure it. Indeed, in the truth which God is, no suffering which befalls is us is so minor, whether it be a kind of discomfort or inconvenience, that it does not touch God infinitely more than ourselves …. [22]

This understanding of God's presence reminds of Julian of Norwich (1342-1423) and her *Revelations*. She saw clearly how deeply Jesus cares about all that is made.

With God so directly involved in a creation that, as I suggest, exists by universal spirit (which is a derivative of

God's Spirit), it must be asked what in essence universal spirit is like. For God will not create other than in accordance with God's own nature. God is Spirit. God is Love.[23] Love is the nature of God. God's Spirit is the Spirit of Love. It follows that the nature of universal spirit is love also. The very essence of love demands that it will seek the relational come what may. The idea that a God of Love deems it fit to remain remote from the divine handiwork, is not biblical. Creation is a derivative of God and it remains fully related to its source; not a fruit that fell from a tree. Fortunately, in theology, the tide is turning towards a modified perception in which God's direct and continuous engagement with creation is accepted.

There is a place for God in our overly mechanistic worldview. Creation exists by the influences of spirit and of love. Love as a first principle in creation needs an explanation though. One that enlarges upon what love is commonly understood to be. Its opposing force called sin likewise needs further thought. (It will be discussed.)

Physics, based on reason, can only ever partly explain the focus of its enquiry – our universe. Christians may avail themselves of a deeper understanding, one that incorporates the spiritual.

Chapter 3
Science and Religion

The question, 'Is anyone out there?' is as old as people walking the earth. They would have perceived the sky to be a canvas across which sun, moon and star moved in predictable ways; determining the rhythm of the days with light and darkness. The earth was the centre of their universe. For years it was considered to be flat. Only in the 3rd century B.C.E, Hellenistic astronomy determined the earth to be a globe. Come the Middle Ages and Copernicus found that the sun was the centre of the universe with the earth moving around it. Today we know that our precious world is far from central in cosmic dynamics. The nature of the sky is interpreted very differently.

Many stars can be seen with the naked eye. Use a telescope and the heavens open up further. What is observed remains only a scratch on the surface of a massive expanse. Most of it will stay beyond visual observation. Visible light cannot cut through the ever-present celestial gas and dust in space to reveal the universe far way. Those discoveries are made with radio waves, which pass through like radar in a fog.[24] From how far away these radio waves, emitted from celestial bodies, may reach the earth, remains unknown. The universe is extremely large with an expected radius of 46 billion light years.[25] It can throw up information stranger than fiction. For instance, its radius does not in any way align with its age. Astrophysicists estimate that to be 'only' 13.7 billion years.[26] Further discoveries are made continually. It shows how much more there is yet to be learned.

The question, 'Is anyone out there?' has a long history and remains important. In the early days of human existence primal consciousness responded with natural simplicity. The insecurities of life, insignificant homo sapience in a world determined by nature and the sky, made a spiritual response essential. The heavens and the natural world needed human appeasement. Religious consciousness was embedded in the cosmos and in the fertility cycles of nature.

That changed in one of the most dynamic times of religious development the world has ever seen - over a period of 600 years from 800-200 B.C.E reaching its peak 500 B.C.E. During those years human consciousness achieved a much increased reflective ability. German philosopher Karl Jaspers called it the

Axial Period. Because, 'it gave birth to everything which, since then, man has been able to be,' theologian Ewert Cousins explains, quoting Jasper.[27] Cousins further writes:

> The Axial Period ushered in a radically new form of consciousness. Whereas primal consciousness was tribal, Axial was individual. "Know thyself" became the watchword of Greece; the Upanishads identified the *atman*, the transcendent centre of the self. The Buddha charted the way of individual enlightenment; the Jewish prophets awakened individual moral responsibility. This sense of individual identity, as distinct from the tribe and from nature, is the most characteristic mark of Axial consciousness. ... The self-reflective *logos* emerged in the Axial Period, it tended to oppose the traditional *mythos*.[28]

Thus humanity evolved with today the modern sciences simply ignoring the question, 'Is anyone out there?' It sees the sky very differently from the first humans. This development disturbs Christianity. The other major religions are largely unconcerned. In addressing why, a brief description of the nature of each religion is helpful.

Hinduism is an ancient belief originating in India that is so diverse, it is impossible to categorise it as a singular faith expression. Its insights are derived from seers and writings that are eclectically adapted to every new era. 'Hinduism is basically an array of techniques for establishing linkages between the human world and the transcendental world beyond it.'[29] It is the oldest of the main religions. Though varied in its worship, depending on tradition, it recognises three terms for an Absolute Reality, which itself is One. These terms, *brahman*, *atman* and *purusha*, are beyond word and thought.

> When you look at the world around you, you see the source of all in *brahman*. …. You look within yourself and you see that the source of your own being and consciousness is the *atman*, and that the *atman* is one with *brahman*. … There is the conception of the whole universe as being a great person, a *purusha* …… This *purusha* is the supreme Person who fills the whole of creation.[30]

Reference to a supreme Person must be understood as *non-relational*. For 'Brahman is the Absolute Reality, unchanging and featureless, self-existent and alone. But *brahman* is also all this world, the phenomenal world of change and manyness.'[31] What is visible in the world exists as invisible in *brahman* and vice versa. With Hinduism the science and religion 'question' does not exist.

Buddhism derived from Hinduism with a major difference. 'The Buddha started from the opposite point of view to the Hindu. He was concerned not with the mystery of being but with the fact of change and becoming.'[32] Buddha considered the world a place of *dukkha* (suffering) and determined that ultimate happiness was to be found in escaping the cycle of birth, death and rebirth. 'When life is described as suffering, it does not mean that pleasurable experiences are not pleasurable. It is simply that they remain subject to change and loss. By comparison with spiritual pleasures they are not satisfying.'[33]

Buddhism differs from other religions in that it does not recognise the person as a concrete 'self.' Identity is an illusion.

> Buddhism stand unique in the history of human thought in denying the existence of such a Soul, Self or *Atman*. According to the teachings of Buddha, the idea of self is an imaginary, false belief which has no corresponding reality, and it produces harmful thoughts of "me" and "mine", selfish desire, craving, attachment, hatred, ill-will,

> conceit, pride, egoism, and other defilement, impurities and problems.[34]

To Buddhism, human achievement has no bearing on its understanding of the divine realm. It can accommodate the sciences with ease.

It might be expected that the advancement of the sciences will trouble all three monotheistic world religions. Islam and Judaism, however, do not have the issues Christianity is grappling with.

Islam does not make a clear distinction between the natural and the supernatural. Nature is a divine portent. It is a book that can be read like the Qur'an, which is a portent itself and the source by which nature's cosmic text can be understood.

> The cosmic text is a primordial revelation whose message is still written on the face of every mountain and tree leaf and is reflected from the light that shines from the sun, the moon, and the stars. But as far as Muslims are concerned, this message can be read only thanks to the message revealed in by the Qur'an.[35]

Insight is based on revelation. Any other approach is

irrelevant to religious understanding, which is the true understanding, and the one that really matters. Science is welcome to its findings.

At the second European Conference on Science and Religion, held in The Netherlands at the University of Twente in 1988, M Bloemendal was the only speaker presenting a Jewish position. From his address the following observations are taken.

Many are of the opinion that science and religion are contradictory, and that the knowledge about nature may lead away from religion, whereas the Jewish tradition suggests the opposite. A true knowledge of nature may strengthen religious belief through the increased admiration of the believer for the greatness of creation. A quotation from the Talmud, in which the expected qualities required of a candidate for election to the Sanhedrin are discussed, reads:

> We don't appoint in the Sanhedrin, whether it is a big one or a small one, but sages excellent in the wisdom of the Torah, and owners of great intellect, people who at least know something about other sciences like medicine, mathematics and astronomy; moreover, something about astrology and idolatry in order to judge them.[36]

Although science is considered to be of importance, it will remain subordinate to the rules of Jewish religion as to its acceptability.

For Christianity it was not merely a matter of acceptability but of power. The Church insisted spiritual matters and their derived authority to be superior to any kind of other knowledge. Historically, for about a millennium, in the demarcation between Church and State, the Church held the upper hand. Until the Reformation in the early 16th century. Though the upholder of the spiritual, the Church promoted intellect to the detriment of the truly spiritual. Dogma instructed how the faith was to be adhered to and society became subject to its decrees. Until in this arena of ideas dogmatics became challenged by science and philosophy. Society had become tired of a Church that was more operational than spiritual. It was dictatorial and corrupt. During the Enlightenment, following on from the Reformation, society was ready for a change and open to the creative developments of the sciences and the liberal arts, both being critical of religion. The extent of that shift was significant and the Church had only itself to blame. The reasoned approach of the sciences outshone the reasoned approach of Church doctrine and it lacked spiritual depth to reply effectively.

The signs of spiritual decay had been recognised centuries before the Enlightenment, but were never acted on. During the 13th- 14th centuries, the Holy Spirit initiated an awakening of mysticism. It was a heightened spiritual response to the displeasure felt by the ordinary Christian about the overly intellectual influences of scholasticism, as taught in medieval European universities (1100 onwards). A church renewal was hoped for, but it never happened because church hierarchy suppressed any deviation from dogmatic expediency. The books of Meister Eckhart (1260-1129) were publically burned. Julian of Norwich (1342 -1423) became an anchorite and allowed herself to be locked up permanently in a cell attached to a small church in Norwich. It still exists. Even the Bishop had no right of entry with an anchorite, so Julian wrote her *Revelations* in peace. John Ruysbroeck (1293-1381), who was influenced by Meister Eckhart, wrote extensively about personal union with God and was watched suspiciously. Deep spirituality was frowned upon by the church and suppressed for it challenged the church's religious power. These three are just a few of the mystics of that period.

Theology could have benefitted from mystical influence. It was recognised by John Gerson (1363-

1429), who became chancellor of the University of Paris in 1395. He insisted on the primacy of mystical theology over and against a merely intellectual approach in formulating doctrine. Copleston explains:

> What he objected to was not scholasticism and philosophy as such but rather the overstepping of limits (logicians deciding metaphysical issues, metaphysicians making incursions into theology) and the contamination of Christian theology by a strong dosage of Greco-Islamic necessitarianism. He was concerned with the preservation of Christian faith in its purity, but he was not an anti-intellectual.[37]

John Gerson understood that mystical theology, these days better called spiritual theology, needed promoting. Unless it occupied a prominent position in theological teaching, theology would remain in an arid and impoverished state. A real deepening of the religious life could not be achieved by divorcing mysticism from theological reflection.[38]

It was all to no avail. The Roman Catholic Church remained spiritually superficial and was caught off guard by the powerful reactions that occurred eventually. Its modus operandi was challenged by Martin Luther (1483-

1546) resulting in the Reformation. It encouraged the secular dynamic of the Enlightenment to follow suit – now challenging Protestantism as well. The liberation of the mind from dogmatic scrutiny was unstoppable. It evolved into what the Western world is like today.

Brief historical accounts are an over-simplification. What shines through though is the problem of dualism – form over matter, reason over whole-of-person experience. Modern society ascribes to form predominantly, as was the case in the Greek polis of old. In this, the influences of the sciences have been instrumental. Science has changed the question, 'Is anyone out there?' into 'What is out there?' Society has become mechanistic accordingly.

Theology has the task ahead of bridging a gap. It must embrace the sciences without compromising its own methods of enquiry. Also, it must take spiritual theology seriously and allow for creativity in revisiting ideas that could be modernised. Flexibility is needed towards a better conceptual interaction with believers and society alike.

Theology must become better acquainted with scientific discovery. Scientist and theologian John Polkinghorne, who is a proponent of natural theology, notes that this discipline is advanced mostly by scientists rather than theologians.[39] Improve understanding of the sciences will not be forthcoming, if theology prefers to concentrate elsewhere. Platitudes like, 'Whatever it looks like, it's God,' are unacceptable. Science is possible, because of God who created the universe with consistent laws. It makes investigation possible. That investigation is achieved by the crown of creation, the human being, who has this God-given ability. Our universe and its scientific interpretation is worthy of theological attention. The sciences must be embraced.

Equally important is the furtherance of the spiritual side of theology. It concerns the religious experience of the whole person. A whole-of-person approach will offer a better perspective on reality. For instance, consider love. Dogmatic theology can conceptualise quite readily on love without having any idea what a deep love actually feels like. Not to mention having any real notion of the Love of God. Defining biblical concepts without actually having a 'feel' for how to interpret these in light of God's nature, as revealed by the saints and mystics for instance, does not result in adequate scriptural understanding. That 'feel' is available to anyone who asks the Holy Spirit and allows the necessary time for inspiration. When relying on dogmatics alone, the Gospel will be shared mainly from a rational perspective. That is unattractive to the modern person. Add the often reported misdemeanour of the Church, and a ritual and cult that is considered medieval and dated, and the battle for engagement has become very difficult.

Christianity need not cease from being dogmatic for it serves to safeguard the faith. But it runs the risk of putting the head before the heart; regulation before compassion, in a society that perceives the Church as dictatorial and inflexible. Not surprisingly, the sciences and philosophy are preferred as acceptable bearers of wisdom. Even though it represents a mechanistic rather than an all-encompassing view of life. It lacks the truly spiritual. The Church must ensure that its influences first originate from the truly spiritual, prior to the intellectual and praxis.

The sciences need no opposing except for the idea that science is all-sufficient and God is dead. Science and religion can be complimentary within the boundaries of their expertise.

Chapter 4

Science and Personhood

'Is anyone out there?' has imprinted itself on human awareness since early history. It was however not until the Axial period that a second important existential question increasingly presented to consciousness. People began to consider, 'Who am I?' It is the most prominent personal question in modern Western culture to which a complete answer is still lacking. People live in the knowledge that they do not really know who they are. They also sense, that they will never fully find out. There are hidden dynamics in play. Personhood is a deep well.

'Who am I?' became central to the development of the major religions. From taking nature at face value and placing the gods directly into that context, during the Axial period the mind began to interpret the heavens philosophically. The notion of divinity began to correspond with the evolving of consciousness.[40] By the understanding of its seers, each major religion acquired unique features. For Hinduism, it meant many gods. Buddhism had none. Nor did Taoism, which is a wisdom philosophy. Non-Eastern religions, those of

Arabic origin, adhered to monotheism. The idea of one supreme God and no other. Western culture derives from that.

The nature of monotheism is a mind focused on responsible behaviour that follows the dictates of God. Quite unlike Hinduism and Buddhism, which do not have an explicit moral code. At first, monotheistic moral behaviour was fully culturally dependent. In Old Testament Judaism, a godly consciousness became instilled through social ceremonies and prohibitions. Christianity, while using these controls, took moral understanding a step further. Behavioural awareness was maintained from within the person. A personally recognised knowledge of morality. It was intrinsically part of being human and not necessarily dependent on social directives. In the monotheistic religions, a believer was held responsible before a relational God. This God could read every person's spirit individually and would evaluate one's life - its motivation and behaviour. This is the Western world's understanding of the idea of God.

St. Augustine (354-430) grew up under the influence of Greek philosophy which possessed a high degree of self-awareness. Once realising that life would never be truly meaningful unless submitting himself to that one God as understood by Christianity, he succumbed. The modifications of his thinking and emotions were written as an autobiography. Krister Stendahl suggests that Augustine's *Confessions* may well be, 'the first great document in history of the introspective consciousness.'[41] It was a step forward in what consciousness perceived itself to be. Augustine's book, which is still frequently referred to today, reached a level of self-awareness that somehow was to dissipate soon after. Around 500 C.E. consciousness began to diminish. Tor Norretranders explains, quoting Morris Berman.

> Human self-awareness, for reasons not entirely clear, seemed to disappear during this time and then mysteriously reappear in the eleventh century. Behaviour during the period A.D. 500-1050 had a kind of "mechanical" and robotic quality to it.[42]

This reawakening began most notably with the English and Rhineland mystics. Richard Rolle (1290-1349) wrote *The Fire of Love*, which was the first book written in the English language. The well-known text, *The Cloud of Unknowing*, originates from the latter half of the 14th century. The mystical movement culminated with the Spanish greats, Teresa of Avila (1515-1582) and John of the Cross (1542-1591). John is considered one of the finest Spanish poets. Based on visions, he drew a picture of Christ on the cross seen from above, at an angle that offered a new perspective to art. The mystics were a God-given gift to the Church in preparing itself for the secular developments ahead. Unfortunately, the opportunity was wasted.

The Renaissance (14th-17th century) took off with a vengeance. The arts, science and technology began to develop at a pace. The Reformation was followed by the Age of Enlightenment (17th-18th century). It was to change culture altogether. Human cognizant ability came out of its slumber to reach heights not previously seen in Europe. This consciousness has reached further levels today. But the corresponding loss of spiritual connectivity puts the true value of this achievement in doubt. Modern consciousness does not encompass all that life is about.

The arrival of psychoanalyses, with Sigmund Freud (1856-1939) probing the subconscious, and the many psychology theories that followed, have been helpful. But plenty of questions about the psyche are left unanswered.

It was only a matter of time before the sciences began to add their insights to the idea of being. The use of quantum in biochemistry opened the doors. Science presents a worldview in which personhood becomes reduced to a manifestation of physics. The prediction is that quantum reality will one day explain all that personhood entails. The mind/body problem will be resolved. Nature will be unveiled as no more, and no less, than an interaction of chemical processes. The question 'Who am I?' is materialistically formulated as, 'What am I?' Materialists hold that the foundation of everything is found in matter.

This scientific reductionism does not align with personal experience and finds no ready acceptance with ordinary people. They sense the question 'Who am I?' to be spiritually valid. But they may not subscribe to a need for, 'Is anyone out there?' The two questions can exist independently of each other. God may be dead, but personhood is not and remains of deeper significance than what the sciences are likely to reveal. Not all scientists are satisfied with an exclusively mechanistic theory. They ascribe to dualism. The idea that the mind, and thus consciousness, exists independently from its physical substrate. The scientific world is divided on the issue with materialism the dominant view.

Philosophy has entered the fray. Thomas Nagel's article, 'What is it like to be a bat?' is in support of dualism. However much we might know of the physiology of a bat, we can never really know what it would be like to be a bat. The objectivity of science cannot penetrate the realm of subjectivity. It is a problem that not science but philosophy might solve one day, David Chalmers suggests. The key to it is information. Information should be considered as essential a property of reality as matter and energy.

The idea of information is not unknown to physics. John Horgan makes a comment about it.

> The concept of information does not make sense unless there is an information processor – whether an amoeba or a particle physicist – that gathers and acts on it. Matter and energy were present at the dawn of creation, but life was not, as far as we know. How, then, can information be as fundamental as matter and energy?[43]

Horgan is writing about science here and not philosophy. In 1988, Thomas Nagel gave a lecture at the Royal Institute of Philosophy in London with the title, 'Conceiving the Impossible and the Mind-Body Problem.' He proves that in using current concepts a philosophical connection between mind and body cannot be logically established. Creative thinking beyond the usual is called for, if a solution is ever to be found. The likelihood of that I consider questionable. In concluding the lecture Nagel explains:

> Truly necessary connection could be revealed only by a new theoretical construction, realist in intention, contextually defined as part of a theory that explained both the familiarly observable phenomenological and the psychological characteristics of these inner events. Its character would have to be inferred from what it was introduced to explain – like the electromagnetic field, gravity, the atomic nucleus, or any other theoretical postulate. This could only be done in the context of a theory containing real laws and not just dispositional definitions, otherwise the theoretical entity would not have independent reality.[44]

Scientist and theologian John Polkinghorne uses the idea of information differently again. He considers it feasible that God's interaction with creation occurs in the form of information input. But it influences creation in non-energetic ways.[45] I suggest, that this occurs as an expression of spirit. With the understanding that spirit is both an influence *and* the energetic source of creation. That this can be explained as envisaged by Thomas

Nagel is doubtful. Still, the creativity of philosophical thinking on the issue is commendable.

In philosophy, the ideas about consciousness vary. John Searle, focussing on neurobiology, compares the occurrence of consciousness with the manner in which molecules construct solidity. A wheel rolling downhill is solid though it consists entirely of molecules. About the brain he suggests: 'Just as the behaviour of the molecules is causally constructive of solidity, so the behaviour of the neurons is causally constitutive of consciousness.'[46] It is an attractive idea for the brain is central in the operation of consciousness. But what is it - that guides the neurons' behaviour and our consequent awareness? Has the feeling of continuity, which everyone experiences, a material foundation? Not so, according to American physicist Richard Feynman.

> The atoms that are in the brain are being replaced; the ones that were there before have gone away. So what is this mind of ours: what are these atoms with consciousness? Last week's potatoes! They now can remember what was going on in my mind a year ago.[47]

Nevertheless, physicists imagine correlation between the operations within quantum states and those of consciousness. Freeman Dyson wrote: 'The processes of human consciousness differ only in degree but not in kind from the processes of choice between quantum states which we call "chance" when made by electrons.'[48]

Suggestions like these are not based on experiment or observation. Rupert Sheldrake informs that,

> In quantum physics, the wave function that describes how electrons and other particles might behave is a mathematical model in a multi-dimensional space based on "complex numbers" that include the imaginary number, the square root of -1. …… The wave function itself is not material: it is a mathematical description of possibilities.'[49]

Sheldrake proposes that the mind engages with perceptual fields.

> The gravitational field of the earth is inside the earth and also stretches out far beyond it, keeping the moon in its orbit. The electromagnetic field of a mobile phone is both inside it and extends all around it. …..I suggest that the fields of the minds are within the brains and extend beyond them.'[50]

He supports this idea with an experience familiar to all. The mind touches what it sees. 'Therefore, I might be able to affect another person just by looking. If I look at someone from behind when she cannot hear me, or see me, and does not know I am there, can see feel my gaze?'[51] Everyone knows that to be possible.

The mind as a field brings to mind the idea of theologian Wolfhart Pannenburg that the Holy Spirit might be a field.[52] This suggestion has been heavily criticised. The Holy Spirit being a Person is best not held captive by field theory. Universal spirit, however, could be imagined as a field quite readily. The description of such a field reminds of Alfred North Whitehead (1861-1947) and process theology. The possibilities are as follows.

From the universal spirit field, fundamental to creation, energy and matter originate to function in accordance with spirit-given information. It ensures the continuity of creation as an ever-developing process. This process unfolds in cinq with the inherent potential and possibilities of each entity which manifests in events. Nature is forever in flux. Spirit information automatically initiates change. This change is the naturally occurring processes. It may be interfered with, if God so decides, with different information from that already functioning. Creation is not pre-scripted and naturally evolves. However, it can receive moments of special guidance. It is the spirit field that makes creation possible and the laws of nature are contained within it. This field is a direct derivative of God's Holy Spirit, with no relational demarcation between the two. Neither are spirit and matter differentiated in their origin.

A dualism between matter and spirit, at the most fundamental level, is non-existent. The 'Who' and the 'What' are inseparable. 'What' originates from 'Who'. They are integrated in a bond of love, which is the primary nature of spirit. This spiritual and existential bond will not be broken because the nature of love prevents it. Every dualism occurs as an existential perception. At creation's primary level demarcations are unsustainable. At that level, all is one.

Science prefers to be non-dualistic also. But in a different way. Many scientists view the deeper nature of reality as old-fashioned superstition. As does philosophy. Within the individual it is never felt as if dynamics of spirit and body operate alongside each other. Because the unity of personhood overcomes the sense of a body/spirit duality. Therefore the sciences see spirit as an illusion. Certainly not a phenomenon that could exists apart from the body. All is material. All is one.

The true nature of personhood is thus under threat. Philosophy will be of little help. An extensive survey has established that only 15% of philosophers hold theistic beliefs.[53] It highlights the importance of theology in upholding the validity of spirit. It must interact creatively with the sciences.

Chapter 5

Spirit and Personhood

Throughout human existence the universe has inspired spirituality. Awe and wonder made people sensitive to the transcendent. The spirits were perceived in the heavens as well as in nature. Science has moved that wonder away from the spiritual to the merely physical. Its explanations about nature and the universe are incredible, often supported by experimental verification. There is much yet to be discovered, but the basics seem clear. Reason and objectivity rule the day.

The idea that life in essence is reasonable and thus objective originates from Greek philosophy. Reason was declared independent from other abilities that comprise a person. However, such 'objectivity' can only ever offer partial insights.

Herman Dooyeweerd saw a complete answer from the sciences about life as impossible. About science and personhood, he observes the following.

> But when one asks them: "what is man himself, in the central unity of his existence, in his selfhood?"

then these sciences have no answer. The reason is that they are bound to the temporal order of our existence.[54]

Dooyeweerd insists that the deepest answers about life are not found by reason alone. Reason is what the sciences have on offer. Dogmatic theology has its limitations likewise it being highly dependent on reason. The most important insights in life come from God directly, Dooyeweerd insists. God speaks to every person equally. Into the heart with a message of saving grace. If science suggests a trust in reason, God asks to trust in knowledge that comes from revelation. Scripture offers information that addresses all of human experience and not just the cognitive.

Nineteenth century German theologian Franz Delitzsch (1813-1890) concluded from a study of Genesis and the human person, that matter and spirit are opposites relatively, but not absolutely.[55] At their primary level of manifestation there need not be a differentiation.

It makes sense, when subscribing to the notion that all existence is upheld by God, spirit as well as matter. Matter, those molecules and energies, under particular circumstances can interact to express the dynamic called life. Then, matter becomes enlivened by spirit. Inspirited matter, in the animated world, may reach levels of consciousness. Its ultimate expression is found in human beings, who are the most intricate of God's creations.

In the story of Genesis, God personally breathed life into Adam, the first person in the bible. The Lord formed man of dust, from the ground, and breathed into his nostrils the breath of life; and man became a living being.[56] Everything living does so by the breath of life. However, there is a subtle difference between the breath given to humanity and that of creatures. Creatures received *ruah*. It means 'wind', or 'physical breath', or 'the Spirit of God', or 'the life force of subhuman creatures.' The word is used in God's announcement to Noah that there will be a destruction of 'all flesh in which there is the breath of life.'[57] In the creation of Adam, the far less common word *nesama* is used for breath of life. An extra dimension was added to human reality, one not available to creatures. The word *nesama* is used for spirit in Proverbs 20:27: 'The spirit of man is the lamp of the Lord, searching all his innermost parts.'[58] The breath of life that Adam received established a relational position towards God, which God will evaluate. The interaction is one of love and holiness – the ability of a personal response to God and moral awareness. John 6:63 states, that it is the spirit that gives life – the body is of no avail. Delitzsch remarks that when, 'God breathes, He breathes forth into the bodily form; and he who breathes, breathes forth from himself.'[59] People are spirit-enlivened and that life comes from God directly in a special way.

But what occurs when that breath departs? The answer given depends on to whom the question is put. Science, being consistent in disregarding the existence of spirit, generally suggests that the 'What' disintegrates into a myriad of independent molecular structures. If there is a level of consciousness involved, it will be no more. This idea of the finality of death is not what many people like to accept. When inescapably confronted with someone close who has died without professing to a faith, platitudes are offered based on soothing ideas that are meant to ease the pain of grief. Always the person still somehow exists rather than having disintegrated.

In Eastern religions the divine is impersonal and not interested in human affairs. Consequently, there is no divine answer given to death. As the world is full of life, which must originate from somewhere, belief is in a rebirth, a reincarnation. In what sort of creature the diseased will find a next life, depends on how well the previous one was lived. This applies to all of nature. All remains on the Wheel of Life.

Monotheistic religions offer a place with God after death. Scripture explains that, 'The dust returns to the earth as it was, and the spirit returns to God who gave it.'[60] The image is one of returning to its origin. But what may that returning be like? Delitzsch gave that some thought.

In his book *A System of Biblical Psychology* (1867), he discussed the existence of the human soul. Till this day his book remains the only significant theological treatise on biblical psychology. He observed that, 'The spirit in man is the source of life ...... The soul, it is true, is also living in itself, but not by itself: it is that which lives in a derived and conditioned manner.'[61] Delitzsch found no biblical basis for the view that at death the soul will cease to exist. However, a certain kind of death to the soul takes place. 'It does not die, so far as it is of the spirit (Matt. X.28); but it dies, so far as it has become of the body. Its life that has emanated from the spirit endures; but its life that is immanent in the body perishes with the body itself.'[62]

When not allowing for the existence of the soul in the conceptualisation of personhood, it may simply be accepted that it is the spirit in which the awareness and experience of bodily existence is recorded. Either way, the point is that upon its returning to God, the spirit is not a blank slate. It carries an imprint of all that a person experienced in bodily form – a personal history. That history includes one's bodily form. From conception spirit and body develop together in accordance with a person's DNA. Spirit and body are inseparable until death. Out-of-body experiences are somehow possible, but the spirit remains anchored to the body. Body affects spirit and vice versa. All is recorded in spirit and will endure.

Philosophically, it may be suggested that there are three dynamics involved in shaping creation. Each dynamic is contained in spirit and concerns 'information'. Firstly, there is the dynamic of form. What will an entity look like, however brief its existence? Secondly, there is function. What is the expression of that entity? Finally, every entity caries the seed of its perfection. It will be realised in God's New Creation that is to come. For people it means that physical imperfection experienced on earth will dissolve with the material body into the dust. Imperfections of spirit are covered in Christ. The New Creation promises a perfecting of the person in every regard. All of creation will thus be made perfect in accordance with God's holiness and not just people.[63] The Gospel story can become overly anthropocentric which does an injustice to the magnificence of God's plan overall.

The idea that a person's history is not erased at death, is finding some traction amongst scientists. Roger Penrose suggests that proteins, which are the structural components of human cells, carry quantum information on a subatomic level. When a person dies, the quantum information probably continues to exist and remains in the universe forever.[64] A former manager of the Max Planck Institute of Physics (unnamed) concludes that thus we are immortal.[65] Quantum information may possibly exist forever, but in a new universe.

The New Creation is a realm different from what came before. Upon its completion every present realm will cease to exist. Our universe and the spiritual realms of angels and demons. The resurrection of Christs covers every aspect of every realm.

Karl Rahner gives penetrating insights in a reflection on Easter. It concerns the Son of God, upon death, descending into the lower regions of the earth from where he arose. In his resurrection, Christ never left the Earth behind, Rahner writes, but carried it in himself to be presented wholly before the Father.

> What we call his resurrection – and unthinkingly take to be his own private destiny – is only the first surface indication that all reality, behind what we usually call experience (which we consider so important), has already changed in the really decisive depth of things. His resurrection is like the first eruption of a volcano which shows that God's fire already burns in the innermost depths of the earth, and that everything shall be brought to a holy glow in his light. He rose to show that this has already begun.[66]

For Rahner, the new creation process has already begun, not only within the believer, but within the earth itself. The new power of transfiguration has taken presence 'in the world's innermost heart.' One day this reality will be revealed in glory.

A New Creation was declared by Jesus when he met with his disciples after his resurrection for the first time. It involved the breath of life of a new dispensation, by a new manifestation of spirit. First, Jesus wished the disciples 'shalom.' It was not this time meant as a common greeting, but signalled the future to come. Walter Brueggemann explains shalom, that magnificent greeting of Israel. 'It is an announcement that God has a vision of how the world shall be and is not yet.'[67]

After his greeting Jesus breathed on the disciples. The similarity to God breathing life into Adam is readily recognised. This time however, it was a 'new life' that Jesus imparted. Adam's life came by universal spirit in which the present creation exists. Not so the breath Jesus blew onto the disciples while saying, 'Receive holy spirit.'[68] It was 'holy spirit' that was imparted. As Jesus had predicted when telling Nicodemus that to enter the kingdom of God one would need a spiritual rebirth.[69]

On purpose I have written the words 'holy spirit' without a definite article or the customary capital letters. In translating the New Testament from Greek into English, an important detail about the operation of the Holy Spirit and the new reality became lost. The introduction of the designation 'the Holy Spirit' into God's history with humanity appears to be particular to the New Testament. The Old Testament refers to the Spirit of God or the Spirit of the Lord rather than the Holy Spirit. Possibly, because the dispensation of the Holy Spirit was yet to be. Jesus told his disciples that unless he departed, the Holy Spirit could not come.[70]

The use of the word Holy Spirit in English is always preceded by its definite article reading: *the* Holy Spirit. In the original Greek that definite article only appears occasionally. It happens when referring to the Holy Spirit either in Person or as the Source of spirit action in the new dispensation. This spirit action itself is referred to as 'holy spirit'. As mentioned, Jesus breathed 'holy spirit' into his disciples. When Paul asked his hearers about how much they knew about the spiritual possibilities open to them, he said in Greek: "Have you received 'holy spirit' when you believed?"[71] The absence of the definite article is important. It allows for the suggestion that 'holy spirit' and universal spirit are similar operationally, but of a different value. Presently, creation exists in universal spirit. One day, at God's command, it will come to exist exclusively of 'holy spirit', without spot or blemish.

Wolfhart Pannenberg, in reference to scientific field theory, suggested that the Holy Spirit might be a field.[72] In the New Creation it would be more accurate to consider holy spirit not a field but a divine presence. (The same might be suggested for universal spirit.) Jesus told his disciples that were two or three are gathered in his name, he would be in the midst of them. He dwells within each believer individually while all believers together around the world form the body of Christ. Whatever the true reality of universal spirit and 'holy spirit', the designation of field as used in scientific understanding is bound to be different. Spirit realities are deeper than that.

When 'holy spirit' enters a person, the dynamics between universal spirit and a material body do not become obsolete. But a significant spiritual change in being has occurred. Primarily, it concerns the relationship with God and assurance of a heavenly future. A believer still very much remains in the world with its joys and sufferings. But is no longer 'of the world'[73] because of 'holy spirit'.[74] That person has entered a New Reality. That reality has current benefits that can be brought into the affairs of every ordinary day. A sense of belonging to God, a heightened spiritual and moral awareness, are some of those. A believer is the first fruit of the New Creation to come. Eventually, all of creation will exist in 'holy spirit' exclusively. The whole universe carries this promise deep within.

The human person is the crown of God's creation. Personhood is immensely wonderful. In the creation spirit finds its ultimate realisation in people. God declared every person to have been created as an *imago dei*. 'Let us make human beings in our image, after our likeness,' (Gen.1:26). The meaning of this verse has been much discussed. Two main views can be distinguished. There is the substantial/structuralist view. People have attributes and capabilities also found in God, e.g. reason and will. The second approach is a relational one. There is a fundamental and potential relationship between God and humanity. This relational ability translates in society as the need for community. Only in community can *imago dei* become sufficiently expressed and will the full realisation of personhood become possible. Theologically, that is in Christian community. Theologian Emil Brunner saw *imago dei* as relational supported by the structural. 'Man's relation to God is not to be understood from the point of reason,[75] but reason is to be understood from the point of view of man's relation to God.' These technical differentiations are interesting. But they are intellectual descriptions that don't do justice to personhood in all its facets and its depth.

There is no question that every person exists as an image of God, for God said so in the first chapter of the Bible. That image in its current state is marred by sin but has not thereby become undetectable. *Imago dei* is best described briefly as people reflecting the nature of God.[76] Presently, it is but a dim reflection with sin as the spoiler. Many human actions are not representative of the nature of God. In the New Creation the effects of sin will have been erased and *Imago die* shines brightly and forever.

Theologian Karl Rahner places the resurrection of Christ in a universal perspective and comments about people.

> Ever since that event, mother earth bears nothing but transformed children. For his resurrection is the beginning of the resurrection of all flesh. Once thing, of course, is necessary for his event – which we can never undo – to become the blessedness of our existence: he must burst forth from the grave of our hearts. He must rise from the core of our being, where he is power and promise.[77]

This is the magnificent promise of spirit to personhood.

Chapter 6
In a world of logic

The progressive liberation of the Western mind, from Renaissance to Enlightenment and beyond, has affected all aspects of human endeavour. Over time the sciences became predominant in changing societies. Their investigations presented a world not ever imagined. Creative imagination and the ability to test its validity with increasingly intricate procedures allowed for the development of knowledge and technology as known today. There is great passion involved in doing science properly. But the impression left with the person in the street is one of a mechanistic kind.

Central to scientific discovery is the ability to discover coherence in that which presents as seemingly random. The manner in which understanding comes about is by insight and logic. Anything that remains a mystery is considered not to have been penetrated logically yet, whether by systematic reasoning or mathematics. In modern society logic has a central role. Its supremacy affects people more than they realise. The very nature of logic excludes much of what life is about overall. The word logic has a sterile feel to it. It is highly effective in gaining reliable outcomes, but it lacks an emotive quality. Technology, which depends on logic, has created the possibility of self-sufficiency for many people. But this sufficiency can be shallow bearing in mind the overall potential of personhood.

David Brooks writes that every year researchers from UCLA survey a nationwide sample of college freshmen to gauge their values and what they want out of life. He compared the findings from 1966 and 1990. The notion of needing a philosophy of life dropped from 80% to less than half that. Becoming rich rose from 42% to 74%.

> In 1966, in other words, students felt it was important to at least present themselves as philosophical and meaning-driven people. By 1990, they no longer felt the need to present themselves that way. They felt it perfectly acceptable to say they were primarily interested in money.[78]

It is an over-simplification to blame logic. And yet, logic is at the centre of this development. With the sciences so prominent in society, and they cannot function without logic, it is no surprise that some scientists of note make us believe it is science that really matters in explaining existence. Philosophy tends to agree. It has given up on dealing with questions of deeper life and turned to logic instead. Two of the most powerful influences in the world, science and philosophy, are focused on the facts of reason. The fruits of logic dominate modern life. In many ways the world has become a better place because of it. Medicine is but one example. In other ways, much is being lost.

The common language that explains our universe is physics. It uses mathematics, the ultimate in logic, as a bridge between reality and theory. Theories are tested as to their sufficiency in representing what is happening in the physical world. If proven correct, it is accepted that nature functions accordingly. Names are given to dynamics and particles under observation. Those names are signifiers. The 'thing in itself', as Kant pointed out, we have no knowledge of. Often we are unable to see 'the thing'. For it's either too small or too far away. Also frequently, the mathematics involved are statistical and include probabilities. However effective, the scientific explanations of reality ultimately are an approximation. What is *really* happening in how the universe functions is unknown.

Science may be compared with tapestry: amazing and a joy to behold. Antique Indonesian cloth presents beautiful intricate patterns very finely woven. Looking at the back of the cloth it is a jumble of yarn criss-crossing all over the place. With the cloth firmly fixed against the wall, I could spent hours interpreting the pattern finding new connections all the time. However, I would have no idea about the yarn that makes the pattern possible. The firmly fixed cloth gives me no access to it.

Scientists are aware of this. Statistical mechanics was first applied by Boltzmann explaining the laws of thermodynamics and became the life's work of Erwin Schrödinger.[79] It has become integral to quantum physics. Classical physics makes use of mathematical models in explaining our universe. Einstein's spacetime concept is such a model. It introduces a fourth dimensions, which cannot be imagined and is not physically real. The idea is a mathematical equation that empirically works in explaining the universe. Mathematical physics is undoubtedly effective. In quantum theory however its creativity has entered a phase with logically impressive ideas that are untestable and lead to the most amazing conclusions. Not all physicists are impressed.

String theory suggests that the universe at its most basic consists of super tiny vibrating strings. It involves further dimensions, potentially as many as eleven. Well-known scientist and presenter Brian Greene is convinced of the validity of string theory. 'But as shall become clear, when seen in its proper context, string theory emerges as a dramatic yet natural outgrowth of the revolutionary discoveries of physics during the past hundred years.'[80] That is Greene's conviction. In using very difficult mathematics, scientists create the existence of dimension upon dimension. Physicist Lee Smolin has reservations about this direction in modern science. 'In fact, neither theory nor experiment offers any evidence at all that extra dimensions exist.'[81] He considers string theory a high risk venture. As with much of quantum theory, its claims are beyond empirical verification and will remain so.

Lee Smolin is of the conviction that all the major breakthroughs in physics have now been made with only further refinements left to be completed. New ideas are needed in addressing some large unresolved issued. One of those problems is that gravity and electromagnetics cannot be combined into one theory. The theories of relativity and quantum each are incomplete also, both having defects that point to the existence of a deeper theory. Smolin suggests:

> The mind calls out for a third theory to unify all of

physics, and for a simple reason. Nature is in an obvious sense "unified." The universe we find ourselves in is interconnected, in that everything interacts with everything. There is no way we can have two theories of nature covering different phenomena, as if one had nothing to do with the other.[82]

Smolin regrets the almost exclusive attention in physics on quantum at the expense of the more classical approaches which are mathematically less open-ended. Quantum is strange in many ways. Science journalist Richard Webb writes:

Quantum Theory is odd, not just because its weird predictions are a source of consternation for physicists and philosophers, but because its mathematical structures bear no obvious connection to the real world, as far as we can see. "We do not have a source for the mathematical formalism of quantum mechanics," says Caslav Bruckner of the University of Vienna in Austria. "We do not have a nice physically plausible set of principles from which to derive it." Quantum physics may be quantum – but as far as we can tell it isn't physics.[83]

Quantum mechanics uses a mathematics that is completely unrelated to any physical manifestation we are familiar with. It moved beyond physicality into a strange reality from which that physicality supposedly originates. At quantum level the laws of nature begin to break down.

Considering the elusive behaviour of particles, how can molecules to hold together at all? It appears that the more atoms in a molecule, the less quantum behaviour becomes observable. Somehow a congregation of atoms 'stabilises' physical processes. It is suspected, that the effect of gravity on larger groups of atoms is instrumental in creating the world as we know it. Probe into the atom itself and beyond, and much is in flux. Why atoms and molecules exist at all, is unknown.

Many of the findings about the infinitely small are odd. Some of these have become well-known, e.g. wave-particle duality. That light presents as a wave or particle depending on how it is measured. Michael Brooks mentions a few more. 'Quantum particles such as atoms and molecules have an uncanny ability to appear in two places at once, spin clockwise and anticlockwise at the same time[84], or instantaneously influence each other when they are half a universe apart.'[85]

That is what physics concludes to date. Perhaps,

the data only seems to give that result. Brooks mentions that, 'any attempt to talk about an electron's location within an atom, for instance, is meaningless without making a measurement of it. Only when we interact with an electron by trying to observe it with a non-quantum, or "classical", device, does it take on any attribute that we would call a physical property and therefore becomes part of reality.'[86]

There is a reality beyond our reality. Physics accepts that, but cannot accept this reality to be anything but essentially physical. It leads to strange possibilities. Classical physics is not consistent and 'is faced with the *uncertainty principle*, which tells us that we cannot measure a particle's position and momentum at the same time.'[87]

Quantum theory is different but even stranger.

> In this picture, wave functions do not "collapse" to classical certainty every time you measure them; reality merely splits into as many parallel worlds as there are measurements possibilities. One of these carries you and the reality you live in away with it. "If you don't admit many-worlds, there is no way to have a coherent picture," says (Lev) Valdman.[88]

It seems that the idea of multiple universes is seemingly necessary in maintaining a consistent physics blueprint.

Physics may well be taking a step too far. For quantum theory is based on statistical predictions of subatomic behaviour - on probabilities. Many experts like Lee Smolin are convinced that, in spite of its success, quantum theory hides something essential about the universe that needs to become known.

Logic has offered a great harvest of insights. Because of its inherent consistency, it allows for theories to develop with a reach as far as mathematics will allow. Logic is attractive, for it excludes any influences that cannot be held subject to logic. It thereby narrows its enquiries and in doing so offers a firm foundation from which to investigate. This is why philosophy has turned towards logic in its deliberations and declared other ways of investigating no longer worthy of interest. Philosophy is seeking to reflect the sciences in the manner of its pursuits. This is particularly so for British philosophers, who began to concentrate on logic and linguistics. The philosophy of the European continent is more poetic and based on human experience. It is humanistic with Marxism as one of its interests.

This shift towards logic is mostly due to philosophy being unsuccessful in answering the major questions of being. Contemporary philosopher Colin McGinn is convinced that 'the great philosophical questions – What is truth? Does free will exist? How can we know anything? – are as unresolved today as they ever were.'[89]

He had an epiphany one day, an intense spiritual experience. It revealed that, 'the great problems of philosophy are real, but they are beyond our cognitive ability. We can pose them, but we cannot solve them – any more than a rat can solve a differential equation.'[90]

Philosophy, by nature, is rational. It observes and interprets. It has narrowed the field to investigating the non-metaphysical convinced that reason cannot be logically effective when attending to the esoteric. One of the greats, Ludwig Wittgenstein, was of that opinion. He set clear boundaries for philosophy. Anthony Kenny writes:

> Wittgenstein was most opposed to the idea that Christianity was reasonable, and that its reasonableness was established by a branch of philosophy called natural theology. Philosophy, he thought, could not give any meaning to life: the best it could provide would be a form of wisdom. But compared with the burning passion of faith,

wisdom is only cold grey ash.[91]

American philosopher William James reflected on the place of reason in a religious context differently. 'It amplifies and defines our faith, and dignifies it and lends it words and plausibility. It hardly ever engenders it; it cannot now secure it.'[92] In other words, thinking about faith originates from the experience of faith. Reason is neither the origin of belief, nor has the final word on it.

For philosophy the existence of God remains on open question. Bertrand Russel concluded: 'I do not myself believe that philosophy can either prove or disprove the truth of religious dogma.'[93] Through the ages, ever since Plato, proof of immortality and the existence of God has been attempted by many - including the giants of theology. Russell observes that, 'In order to make their proofs valid, they have had to falsify logic, to make mathematics mystical, and to pretend that deep-seated prejudices were heaven-sent intuitions.'[94] That is a penetrating critique.

Both Bertrand Russel and William James subscribed to 'neutral monism'. 'The metaphysical notion that everything in the universe is of the same "stuff," whether it is matter or consciousness.'[95] Both mind and matter are reducible to a single more fundamental principle that is neutral between them. Russell accepted the validity of metaphysics, but felt it best left alone philosophically. Logic and metaphysics are not currently congruent.

The "stuff" of neutral monism, I suggest to consist of spirit. Science and modern philosophy could not possibly entertain this idea for it does not fit into their modus operandi. Reason and logic will never definitively prove God's existence. That kind of proof can only be obtained by perception.[96] People, created in God's image, will detect the presence of God only when using the full ability of that image. The Divine Image, which always is fully operative, is responsive to its derivative. God can only be known by the 'whole' person. Central to that communication is spirit, for a person is a spirit-enlivened body. The acceptance of God's existence will never be proven merely by the intellect alone. It is an existential impossibility.

In explaining our universe physics speaks the language of mathematics. Religion, however, does so by using the language of revelation. The first concerns logic only and is rational. The second addresses the whole of life as understood by spirit. These languages are very different in their conceptualisation. If any bridge is to be build, it will need to be philosophical. Merely presenting a revelatory insight in response to a scientific one is unsatisfactory. But improved communication between science and religion should be possible. If theologians are willing to give it the required attention.

A well-known theologian who took the sciences seriously is T.F. Torrance. Others have followed in his footsteps.[97] Torrance, in reference to physicist John Wheeler's idea that the way forward for physics must be through discovering a 'regulating principle' that guides disorder into order, offers an exposition and concludes:

> It may be submitted on the part of theological science, however, that the ultimate regulating principle which he seeks is none other than the *Word of God*, from which and by which the beautiful intelligibility of the universe has been conferred upon, but which like the music of creation, is to be apprehended by hearing.[98]

Torrance considered an interaction between science and Christianity to be important. In the preface to *Christian Theology and Scientific Culture* he wrote:

> It is through deep-going dialogue with science and submission of our own theological conceptions to

the critical questions it addresses to us that we are helped to purge our minds of pseudo-theological as well as pseudo-scientific notions, and so are enabled to build up theological knowledge in a positive way on its own proper ground: God's self-revelation and self-communication to us in the incarnation of his eternal Word in Jesus Christ.[99]

Torrance's answers Wheeler's search for a regulating principle in our universe. It is the Word of God. This is unlikely to foster a meaningful dialogue with the scientific community. It will resonate with Christians well enough, but achieves little. The idea *Word of God* without extrapolation has no meaning scientifically, nor does it philosophically. Torrance uses the term theological science. Ideally, that should refer to ideas that can be theologically justified while having scientific relevance. Or, it may simply mean giving theological answers to scientific ideas while remaining mostly within the conversations of theology. It might satisfy theologians, but is of little help to Christian scientists who are seeking Scriptural relevance in their work. Nor to the ordinary Christian, who would be helped by being offered at least some thought on how God may be actively involved in physical reality as explained by science.

The sciences are influential and people are becoming inattentive to the deeper dynamics of life. A world of logic, attractive in its technological comforts, offers a superficial existence. With the Church losing traction in a society dominated by logic, theology must not shun creativity. On the back cover of his book, *The Contagion of Jesus*, Sebastian Moore makes an informative observation. He describes his book as 'passionate rather than rigorous theology', based on a loving God, a saving Christ, and a church of friendship and discipleship.[100]

Spiritual theology is increasingly popular amongst believers and is the bridge between Church and modern society. It must communicate its ideas with language that accepts the logical as worthy, but the spiritual as essential. A language devoid of esoteric and technical terms.

Our world is searching for the spiritual. The interest in meditation and mindfulness are proof. Christianity though is seen as restrictive and outdated. As taking over one's life rather that liberating it. That perspective needs changing.

The Gospel is an everyday story of simple meaning that has much on offer. It is a logical story laden with spiritual significance. It is non-cultural by nature. Let's learn to tell it well.

Chapter 7

Space and Time

Apostle Paul wrote that our amazing creation came forth ex nihilo – out of nothing[101]. That makes sense for with creation issuing from God it would not have existed before that creative act. Creation is not divine and very different from God. So from what did it come about? Matter, as we know it, did not exists. And thus, Paul's idea of ex nihilo. That God spoke creation into being is recorded at the very beginning of the Bible in Genesis. The New Testament informs that it is held together in Christ, who is called the Word of God. Without a physical beginning creation came to be. God spoke, and it was there.

For the universe to originate out of nothing is a scientific impossibility. Science holds that originally all was condensed to the size of a small pea. Where that pea came from is unknown. A massive explosion occurred, super large and incredibly fast. This Big Bang was the beginning of space and time. Space, because gases expanded and particles became distant from each other.[102] Time because, however fast, that expansion did not happen instantly. Philosophers, including St Augustine, have argued that time existed before the world began. Current scientific consensus disagrees but nobody knows for sure.

It is a certainty that neither space nor time are material. Their reality presents because of movement – of progression. German philosopher Immanuel Kant (1724-1804) made a study of space and time. He suggested that, 'Space is the form of the outer senses, and time the form of the inner sense. Space and time are not entities in the world to be discovered by the mind: they are the pattern into which the senses mould experience.'[103] Kant's philosophy took human experience as its starting point. He recognised other non-material influences that form experience without originating in the intellect, e.g. morality. It is undeniable that space and time are noticeably present to everyone and inescapably so. Space and time are experienced separately but are intrinsically related. They are the conceptual boundaries of our enquiries. Reality always occurs within space and time.

Time is confusing. Book 11 (13-16) of the *Confessions* of Saint Augustine presents a celebrated philosophical discussion about time. St. Augustine questions what it means for God to be timeless and the nature of time as experienced by people. He finds no easy answer. "I know well enough what it is,' he writes, 'provided nobody asks me; but if I am asked what it is and try to explain, I am baffled.'[104]

Time is problematic because of the elusiveness of finding the 'this is it' moment. Whatever the moment in time, it has either passed already or is yet to come, but it is never the 'exact moment'. There is never a unit of duration that cannot be further divided and thus become time itself in its singularity. St. Augustine ends his reflection with: 'The conclusion is that we can be aware of time and measure it only while it is passing. Once it has passed, it no longer is, and therefore cannot be measured.'[105]

French philosopher Henri Bergson in *Matter and Memory* (1896) writes:

> Pure perception, in fact, however rapid we suppose it to be, occupies a certain depth of duration, so that our successive perceptions are never the real moments of things, as we have hitherto supposed, but are moments of our consciousness.[106]

Bergson suggests that when looking at something

we will always see it in the manner our consciousness constructs impressions. That can never be in 'real' time.

The nature of human experience makes 'real' time impossible. Experience involves myriads of impression from our spirit and senses. Using the idea of a 'notion' is helpful in explaining what essentially it is about. A notion is an extremely brief subliminal sensation. Notions are continuous and need not involve awareness. Tor Norretranders explains:

> The fact is that every single second, millions of bits of information flood in through our senses. But our consciousness processes only perhaps forty bits a second – at most.[107]

Awareness sets in when notions reach the level of consciousness, where they may be conceptualised, or not. When conceptualised, they become a perception. Because reality is always changing and notions accordingly, the perception of time occurs. But the first moment of a notion has passed before awareness sets in. We cannot consciously experience that moment.

The rhythms of life are set in time. The most noticeable is the movement of the sun each day. For millennia it determined how time was progressing. Years were measured in seasons and by the position of the sun in the sky – longer light in summer and shorter in winter. That position also determined the time of day and possible personal commitments. When the sky was clouded coming to work on time, or for an appointment, was a guess. Once the clock was invented, every town built a clock tower to regulate the community in being on time for that was economically expedient. Days became divided in sections called hours. The idea of hour was known before but never set out as a measurement of sixty concise minutes. It significantly changed the rhythm of life. Clock-time is an artificial measurement, a frame of reference. The great accuracy of today's atomic clock does not change that fact. Clock-time is not 'real' time.

Space likewise is a continuum of notions that reach awareness. Space involves distance and the possibility of movement. Distance, because a separation is detected between objects, the perceiver being one of those. Movement, because that distance allows for objects to travel. Space is always experienced three dimensionally, as an entity that in itself has no substance. I have learned to interpret the nature of space mentally. The moon may seem close by sometimes, but I know it to be far away – always at a similar distance. As with time, space became measured. In its simplest form that would be in centimetres or inches. Three dimensional space is mathematically expressed in geometry. Time did not fit into that mould until Einstein completed his theory of general relativity. Time is more elusive than space.

Scientist Julian Barber argues that time is an illusion. After many years of study, he wrote a theory known as shape dynamics in which space and time is nothing but a system of relations.[108] We may experience space and time, but actually it does not exist.

Einstein introduced the concept of spacetime with significant consequences for cosmology. In the theory of general relativity, the geometry of spacetime describes the effects of gravity. It is a mathematical representation of the gravitational field. Einstein predicted that gravity would bend light rays and by how much. This was proven correct by experiment when observing the rim of the sun during an eclipse. Light from stars that were behind the sun was visible bending itself around it. The mass of the earth likewise makes the gravitational field flex like a trampoline bent by a heavy stone.

Einstein's theory of spacetime as a reflection of reality has its problems. Mathematically, three dimensions of space and one of time are combined into a single entity spacetime. However, in this mathematical model time is stationary – it does not progress. Time has become 'frozen'. Spacetime is like interpreting reality by taking mathematical snapshots.

Julian Barber considered time an illusion. Inspired by Barber's ideas, Lee Smolin wrote *Time Reborn* (2014) seeking to liberate time from its static state, but without much success. Mathematician and philosopher Tim Maudin decided that, despite the efforts of Smolin and others, their theories still suggest time as stationary, while people clearly perceive it as flowing.[109]

At quantum levels the reality of space and time takes a twist. Quantum particles such as photons and electrons are not bound by the arrow of time. Justin Mullins writes:

> The mathematics of quantum theory says that the quantum state that describes them evolves both forwards and backwards in time. This odd state of affairs has led to some researchers claiming that the normal rules of causality don't apply, so things that happen in a quantum particle's future will affect its past. …. The past state of a quantum particle has no more reality than its future state. [110]

Another discovery in quantum mechanics that defies conventional time is entanglement. Particles, or groups of particles, influence each other instantaneously over large distances. Bring change to bear on one and the other reacts at that very moment with a corresponding change. Being instantaneous, or close to it as has been suggested, this reaction is faster than light. According to Einstein that is impossible. Entanglement has been experimentally verified. The phenomenon, together with superposition – that quantum particles can exists in two locations or physical states at once - is fundamental to quantum computing.

Ordinary computers function on a yes/no choice, on binary bits. Entangled quantum bits, or qubits, use particles that can take on many states, far more than the two of binary computing. The objective is to transfer the information of one qubit to another instantaneously and in doing so carry much more information than is possible in a binary system. It will result in smaller computer chips made of ordinary materials that can hold an incredible amount of data and are superfast. This technology is progressing fast.

> Just a few hundred qubits can calculate a tableau of outcomes that exceeds the number of particles in the universe. So far scientists have created small quantum-computing systems in many laboratories that use up to 10 qubits.[111]

This was written a year before IBM announced in November 2017 that it had achieved a 50 qubits system.[112] The number of qubits is very much higher now.

Our amazing universe is an expression of energy that at the level of the very small does not obey the laws of classical physics. Extremely tiny bits of energy are popping in and out of existence continually. Perhaps they are shifting between universes. It depends on what you believe possible. Time, at those levels, must be different. If in such quantum events time exists at all. Whether it does or not is unknown.

In ordinary life, time is real enough; as is space. But it remains a mystery. Time flows and space extends. Neither have a substance. Both manifest in human consciousness involving the whole person. Memory presents time past. Expectation time future. Both are imaginary. Experience is time now. And a sense of space is always present. It is the sense of time that tends to mark my day. This sense experience is far from consistent however. Time may flow, but in my perceptions it does not do so evenly. Clock-time gives the impression of steadiness, which is a mental adjustment to what my actual feeling of time is like. The experience of time fluctuates depending on biorhythms. When excited and engaged it flows fast. When anxious or bored, it progresses slowly. When asleep, it disappears from my awareness for a good while; I have lost the sense of time. Which is confirmed when looking at the clock in the middle of the night and finding it reading quite differently from what I thought. The sense of time and space is a whole-of-person experience, but the awareness of it involves consciousness. When unconscious, all sense of time and space disappears.

Might it be possible for a person to step out of space and time? Mystical experience makes this happen. Even then a sense of spacial awareness and progression is inserted when describing the experience. Accounts by Old Testament prophets confirm this. Apostle Paul spoke of being caught up to the third heaven, whether in or out of the body he could not tell.[113] The Book of Revelations gives an unavoidable impression of space and time. In which a New Jerusalem is promised with streets of gold.[114] Without that city existing in space and time this image makes no sense. Space and time apparently are always involved within the creations of God. But opinions on this differ and we cannot know.

Space and time are also relative and dependent on

my vantage point. Einstein suggested that to be so. But what if I could step outside of those boundaries and influence space and time, change them somehow? If I could look from the outside in, and make time move twice as fast in creation. What would happen? It would make no difference to how people within the universe experience time. Everything will go faster – nothing excluded. Any person's biological system would have sped up like that of everyone else. So there is no difference to be detected by anyone. The time relations between all that the universe entails would not have changed and people would experience time just like before. Likewise with space: shorten it by half everywhere and human perception of it remains just as it was. The distance between two objects in relation others will be seen as unchanged for the observer would have become half the size of before.

God exists beyond creation – looks from the outside in, so to speak. Time, for God, has no measure. Neither has space. That is why Apostle Peter could declare that one day is like a thousand years, and vice versa.[115] It was a figure of speech. A billion years could be one day, for that matter. For God the sun does not rise and go down to make 24 hours. The size of the universe is irrelevant to God. By placing space and time into creation, God made creation progressive. It brought the possibility of experience, which originates in spirit.

The progression of creation occurs in accordance with the dynamics of universal spirit. It translates into experience for people and those creatures who have a level of consciousness. But everything in nature is subject to the rhythm of time. Spring brings new growth. Winter is the season of rest. Life takes its time. Death brings a stepping out of time as we know of it on earth. Time flows and space extends. Science interprets it one way, personal experience another. Whatever the explanation, it is real enough – to me.

Chapter 8

The nature of spirit

God *is* Spirit. God *is* Love. This is the nature of God. The abilities of God are an expression of Spirit and Love. Every expression of God is a Spirit and Love expression including creation. Creation consists of spirit and dwells in love. All the major religions have discovered that love is the primary reality behind the natural. Love is not reasonable, but it is open to reason. Reason though will never fathom love, as it cannot its Giver. The closest love can be understood is from the heart. The deepest experiences of the heart occur in mysticism. It offers insights like no other.

Every major religion has a rich mystical tradition. It has gained knowledge about the divine that is not obtained through reason. In the last sentence of his book the *Mystical Ark*, Richard of St. Victor (12th century) observes that, '… the depth of hidden things goes beyond human reason.'[116] This depth is experienced as utterly true. St. John of the Cross (16th century) recalls how the experience of God's special knowledge is so convincing that any doubt about its validity evaporates.[117] The uninitiated in spiritual matters cannot understand the nature of that experience. St John explained that revealed knowledge is imparted – not acquired - and it is higher than ordinary knowledge. It derives from a reality that extends beyond universal reality, which functions in relation to it.

But entering into extended reality is not without its cognitive component. The best mystics, C.S. Lewis writes, are little interested in sensory phenomena and visions.

> What they seek and get is, I believe, a kind of direct experience of God, immediate as a taste or colour. There is no *reasoning* in it, but many would say that it is an experience of the intellect – the reason resting in the enjoyment of its object …[118]

Mystical knowledge is not gained by reasoning based on fact or recorded revelation. It has a deeper quality. If the mystics discovered from experience that our universe is held together by love, it must be noted. After all, Scripture tells us that the universe is made through Christs. It exists in him and is held together in him.[119] In Christ, who *is* love. All creative expressions of God are those of love. God's creative acts are spiritual acts. In creation, spirit is a derivative of God's Spirit. It is universal spirit and its nature is love. A universal love that derives from God's eternal Love. Everything exists by spirit and love in a myriad of manifestations.

A pressing question is whether God has created matter spiritless. After all, like love, spirit is non-substantial and without material quality. Quantum theory has shown that essentially matter is a bundle of energy. At quantum levels energy, like spirit, is non-substantial. Therefore, a connection between spirit and energy may be imagined. It will never be scientifically proven but neither will many of quantum's findings. That energy in our universe might originate from spirit is not an unreasonable assumption. Certain answers science is looking for may indicate the presence of spirit. Physicist John Wheeler is seeking a 'regulating principle' in nature. String theory suggest multiple universes to make its theory coherent. Accepting the influences of spirit as a regulating principle that offers coherence would be an elegant solution. But this idea cannot be scientifically developed. It is where the sciences are coming up short in explaining ultimate reality.

If creation originates from God and holds together in Christ, but it has a material nature that is unlike God who is non-substantial, does matter then exist separate from God? This question aligns with the idea that God is 'Wholly Other' (Rudolf Otto) or the Unmoved Mover (Aristotle). God is immaterial and way beyond the mundane. The thought is that the perfection of the divine realm makes it impossible for a struggling creation to exist within its full embrace. But the Gospel indicates otherwise. The Son of God emptied himself and became subject to our creation. Divinity took on human form and became limited to those restrictions.[120] The divine and universal became united in Christ while he walked the earth. The divine is fully capable of indwelling the material. A born-again Christian is proof of it. That person has received 'holy spirit', which is unblemished and the promise of perfection. No duality is experienced in having both a holy and creation bound spirit. Divine perfection and human imperfection can coexist. Divine perfection and material imperfection likewise. The works of spirit and love are never separated from God in whatever form that work manifests.

From spirit all existence originates and the nature of that existence is love. The manifestation of love depends on the entity that is created. At the level of ordinary matter it is undetectable. At the level of plant-life, God's love is easily seen in the beauty of nature.

Animal life allows for a noticeable observation of love. Affection is on display in many species. In the human person, created in the image of God, love most clearly represents what God's love is about even though in a greatly diminished capacity. From spirit, life originates as an expression of love.

God, who is love, is also good. In Greek philosophy Plato recognised the 'Form of the Good' as the highest of all forms. Forms are ideas that represent reality. For every manifestation in nature there is a pre-existing imaginary form. However, the idea of the good cannot be understood solely by imagination. Something more is required. The best philosophy is only possible, Plato advised, if a philosopher gets an intuitive sense of the good rather than only mentally reflecting on its qualities. Plato's 'Form of the Good' may be compared to the influences of love, though he never suggested such. Every good in life is an expression of universal love.

In the nineteen-fifties, psychologist Harry Harlow wrote his now famous paper, 'The Nature of Love', based on experiments with rhesus monkeys. From the observation of baby monkeys, he discovered the deep need for care and comfort that must be met for a healthy development to become possible. Some concluded that Harlow stated the obvious. Perhaps. But during a time in which behavioural psychology and the study of rats was the norm, Harlow's research led to a necessary focus on what is most important to people. The availability of love and care. Harlow was of the conviction that, 'So far as love and affection is concerned, psychologists have failed in this mission.'[121]

His topic of research may have been love, but his methods were brutal. He achieved his findings by depriving young monkeys and making them suffer badly for a lack of nurture and affection. For good reasons he became eventually a focus of the animal liberation movement.

A decade earlier, Hungarian psychiatrist Rene Spitz had observed the equivalent. He compared human babies raised in two institutional settings. The first a foundling home, clean and orderly but a little clinical. The second was a prison nursery, full of rough-and-tumble with lots of physical contact. Tom Butler-Bowdon reports:

> 'Within a two-year period, over a third of the kinds in the foundling home had died, whereas five years later all the prison nursery children were alive. ...... What made all the difference was that the

> nursery kid's mothers were allowed to care for them, while the foundling home's children lived under a controlled regime run by professional nurses. Whether you define "death" as physical or psychological, it was the lack of physical affection and love that was the cause.'[122]

Love is still not at the forefront of psychological enquiry today. It remains elusive – not suited to ready classification.

Philosophy also mostly ignores the topic. Theologian Sebastian Moore laments that a philosophy and ethic centred on love has never been written.[123] Over the years, love has been a topic for reflection, but never from the perspective of its centrality in being human. Kierkegaard, writing about love, transformed it into the highest levels of subjective experience. While Arthur Schopenhauer offered an anti-love philosophy of despair. Frenchman Alain Badiou, one of the few contemporary philosophers willing to address the topic, tells that approaches on love taken by philosophers are threefold. As novelist and playwright he is more readily confronted with love as a subject of interest.

First, there is the romantic interpretation that

> focuses on the ecstasy of the encounter. Secondly, ….. a legalistic perspective ….. love in the end must be a contract …… the system of mutual benefits, etc. Finally, there is the sceptical interpretation that turns love into an illusion.[124]

Badiou himself considers love an existential project that constructs a world from a decentred point of view.[125] It reminds of I-Thou, but not quite. From ecstasy to illusion, from the heights of affection to technique, and all shades in between, is how love has been described.

Harlow's research shows that the presence or absence of love has consequences. It shows that feelings matter for good health. 80% of illness is considered psychosomatic. A troubled psyche leads to ill health. Medical practice, which is applied science, tended to ignore this core issue in feeling unwell until recently. Neurological research, the most advanced of the medical sciences, over time expects to fully unveil how the dynamics of disease are basically chemical. Taking a purely body-focused perspective such a conclusion is inevitable. The spiritual side, however, plays a significant role. With the word spiritual used in a broader sense than what is called spirituality.

Spirit and body interact constantly with regard to health. I may become physically ill, because my body malfunctions. Perhaps because of a natural defect or catching a virus. I can also fall ill due to events that are completely non-physical. Like a death in the family and grief. Depression, though possibly occurring because of chemical imbalances in the brain, frequently has a psychosomatic origin. Whatever the responses of my body under such circumstances, it happens because of a feeling situation. The unsubstantial affects the physical. How that is possible remains unexplained in physics. Psychosomatic illness confirms the influence of spirit upon matter, which is the human body.

The deeply falling in love of two people, its passion and desires, comes closest to what divine love is about in everyday living. Only the mystics will get closer. Whatever the level of experience, it remains a pale reflection of the enormity of God's love. Even so, the human ability to love is attractive. True love has that transcendent feel and people have been fascinated by it for ages. It is best presented in poetry, novels, plays and films. A factual description of love tends to be overly rational and lacks situational background. David Brooks though manages to present an excellent one of a few pages long.

> It is Aphrodite or Cupid. Love is described as a delicious madness, a raging fire, a heavenly frenzy. We don't build love; we *fall* in love, out of control. It is both primordial and also something distinctly own, thrilling and terrifying, this galvanic force that we cannot plan, schedule, or determine.[126]

Brooks continues in discussing a multitude of feeling components in relation to being in love. Love is the deepest and most enlarging of human experiences. It comes closest to what *imago dei* really is about. As Apostle Paul declared: if I have not love, I'm like a cracked bell. He gives his famous description of love in 1 Cor. 13.

Falling in love, this pinnacle of experience, defines love in general understanding. Love is, however, so much else. As is expressed in daily communications. 'I would love to go to the ballgame,' speaks of desire. When someone says, 'I love to play ball,' it expresses the enjoyment of doing so. 'He surely loves himself,' is a comment of excessive self-appreciation. Tears at a funeral reveal love as an emotional attachment. A component of affection is present in all these expressions of love. In his book *The Four Loves* C.S. Lewis discusses four different types of love found in NT Koine Greek: eros, storge, agape and philio – passion, familial, godly and friendship.[127] All of these are expressions of affectivity. But need love necessarily be motivated by affectivity for it to qualify as love?

The key word associated with love is care. Most people believe, that caring without affection is not love. But they are wrong. A person may indeed care with noticeable affection and that will be caring in love. Or, the care may be supported by a sense of relational belonging, in which love may be an undercurrent of affections that are mostly subconscious. It also classifies as caring in love. Furthermore, there is care that just seeks to do good. There are no affective emotions involved. Jesus instructed us to love our enemies.[128] That will hardly include positive feelings, but it seeks to wish them no harm.

The greatest commandment: 'Love God with all your heart, and your neighbour as yourself,' is not suggesting peek experiences. Rather, it addresses our deepest motivations. The commandment might be understood as: 'Care about God deeply and likewise your neighbour as you should care about yourself.' That is the law of love.

All that is good and positive, is an expression of the nature of universal spirit, which is love. All that has beauty presents the love of God in creation. There is great good and much beauty to be found once looking out for it. Life in all its forms originates in love. All of nature has God's absolute commitment for Love can do no other. This commitment will culminate in creation's future perfection. Offering the only Son as a sacrifice guarantees it will come about. God's love is enormous. We may get a glimpse of it by experiencing the heights of human love, which allows an insight into what is to come: life in a New Creation when all is love, without exception. The nature of spirit is love!

Chapter 9

A fearsome power

Our universe exists in universal spirit, the nature of which is love. That may be fundamentally true, but human experience tells us otherwise. Love is well noticeable but so is the destructive power called sin. The whole of creation is subject to its devastating influences and is in bondage to decay. The pressing question so often put is: how a good God can allow that to happen? We will never know, beyond the fact that creation is in this state with a future promise. It will be liberated once God starts revealing a new creation.[129] Scripture must ever be interpreted from that perspective. Focusing exclusively on the here and now does God an injustice and limits biblical interpretation. 'All shall be well, and all manner of things shall be well,' Julian of Norwich wrote. She used this promise many times in writing her book *Revelations*. Unfortunately somehow, in the divine plan, sin is unavoidable.

Jesus made it very clear to Julian when telling her that, 'Sin is necessary.'[130] She was discouraged from asking any further and understood that sin is nothing. It has no substance.[131] But it has legitimacy before God. The sacrifice of the Son makes that abundantly clear. Not that God created sin for it would deny the very nature of God. For whatever reason, when God created, sin infiltrated God's handiwork. Perhaps the divine perfection, the Trinity, exists over and against sin, which lays dormant until it can attack a derivative of that perfection – the creation. Perhaps God's holiness is such because it could be otherwise. It exists against its opposite. We will never know. It is pure speculation seeking to make rational sense of a currently bad situation. A situation that, before any creative act of God took place, was ordained to be redeemed by Christ.

Sin has no substance, Lady Julian understood. Sin is able to infiltrate universal spirit. It feeds on love like a destructive virus and effectively so. It affects every aspect of creation and manifests as disintegration and death. Everything in our universe is temporary. Even the universe itself is, as science confirms. Apostle Peter declared it so many years ago. By the decree of God, 'the heavens and the earth that now exist have been stored up for fire.' Peter foretells that, 'the heavens will pass away with a loud noise.' [132] What exactly this refers to isn't known.

Because sin has infiltrated universal spirit and the

power of love, our universe exists as integration and disintegration. Love put together and perfects. The fearsome power of sin, operating at the same primary level as love, destroys. In cosmology those forces are incredible. The most destructive and powerful are black holes.

Black holes are called black, because nothing escapes from it once sucked in, not even light. They have become significant in explaining the universe, which is full of them. At the centre of our galaxy, the Milky Way, a massive black hole is expected. Apparently, our galaxy features many others. One is expected to exist relatively closely to our earth. Black holes are difficult to locate. They are detected by cosmic waves of electromagnetic radiation. Or rather, the lack of those where a black hole exists. At the rim of it a little light may yet escape. It is a pointer of a possible black hole. It spins at an incredible velocity and absorbs gravity. New Zealander Roy Kerr mathematically proved the existence of these areas in space. What mathematics can achieve is amazing with symbols on a page. Indian physicist Subrahmanyan Chandrasekhar's reaction to this accomplishment is recorded by Marcia Bartusiak.

> Chandrasekhar called that discovery "the most shattering experience" of his scientific life, the realization that Kerr's solution "provides the absolutely exact representation of untold numbers of massive black holes that populate the universe …. that a discovery motivated by a search after the beautiful in mathematics should find its exact replica in Nature.[133]

Massive stars can disappear inside a black hole forever. The death of stars means their complete annihilation. The cosmos is fiercely at war within itself. Composition and decomposition; that is the story of our universe. It involves existence and non-existence, life and death. Things exist because of the creative Love of God. But they are temporal. Everything in our universe, however majestic or beautiful, falls apart either quickly or over a long period of time; and at all stages in between.

On earth the destructive nature of sin is readily detected. In nature sickness and death are the most noticeable. Nature can be very cruel with most species not dying naturally. The idea that sin is limited to morality is incorrect. Its influences are as wide as those of universal love. It infiltrates every aspect of existence. It is only at the level of human consciousness with love and sin being at loggerheads that this translates into morality. At every other level of creation sin is equally active. Life and all that is good originates from universal love. From sin issues death and disintegration.

Moral acts that are purposefully hurtful are called evil. Lesser behavioural imperfections are just sinful. It is impossible to escape from sin. The whole of creation, every aspect of it, is under its power. And under the greater power of love.

Science cannot accommodate such an idea. It needs a regulating agent that shapes the universe but will not consider spirit. Scientist/theologian John Polkinghorne suggests 'information' to be needed in guiding how the universe functions. Spirit, as the primary influence that involves both the dynamics of love and sin, might qualify. It would be of help to Christian scientists who wrestle in understanding why nature originating from a good God is so destructiveness.

Once, I attended a lecture on DNA as part of a science and religion presentation. The scientist explained how jumping genes were instrumental in causing cancer. It was an engaging talk and sometimes technical.

Eventually, our presenter raised the question of why these genes might be jumping and produce ill health. How did it fit the picture of a good God? He had struggled with an answer. Reading a distinguished theologian offered him some insight. The suggested solution was that God allowed free will within creation.

The terms 'free will' and 'free agency' are much debated concepts in theology. It concerns human behaviour and destiny. How free are people in shaping their lives under the power of sin and God's saving grace? To what extent is their eternal future predetermined, if at all? The scientist would not have had this in mind, but merely that somehow creation was allowed by God to operate freely as it saw fit. With genes jumping even when it would have negative consequences like cancer. Whether the concept of will can be applied at molecular level is questionable. Will involves consciousness and is not relevant to physics. The scientist seemed not quite satisfied with the free will suggestion. But knew of no better answer. This is unfortunate.

Creation unfolds progressively in accordance with the forces within and how they interact. Cause and effect are its dynamics. Outcomes can vary. At the level of life this becomes noticeable as bringing health or diminishing it. The former aligns with God's love, the latter with the destructive powers of sin. The misbehaving of jumping genes causing sickness occurs because of the influences of sin at the most primary level. This idea makes sense and theology might explore its potential. A power as important as sin needs further attention. It may be more relevant to understanding reality than imagined.

Christianity teaches that sin's presence is not limited to our physical universe. It concerns the age-old question of whether we are alone in the cosmos. These days the answer focusses on UFOs. In earlier times, when gazing into the starry skies, it would have been about creatures from the world of spirits. Biblically, that is still pertinent.

When asking, 'Are we alone in creation?' then the answer will be: 'Surely not.' Religious tradition is clear about that. Our reality exists within a transcendent one. Religions have variously perceived it. Hinduism, for instance, teaches the existence of a physical world of nature and universe; a psychological one of the senses; a psychic one of angels, demons, evil spirits and saints; and a divine world, that is out of reach.[134] Christian understanding is less nuanced. In addition to the universe it recognises a spirit world of angels and demons, plus the divine realm of the Trinity. The spirit world is a created world that holds together in Christ also, like the natural world. Nothing exists apart from Christ.

The spirit from which the realm of angels and demons issues likewise contains the primary dynamics of love and sin – good and evil. According to Scripture, sin tempted Lucifer, God's highest angelic creation, who became Satan.[135] A host of angels supported his cause and ended up being devils. Their influence on earth is considerable. Apostle Paul clearly recognised the dangers facing the believer.

> Put on the whole armour of God, that you may be able to stand against the whiles of the devil. For we are not contending against flesh and blood, but against the principalities, against the powers, against the world rulers of this present darkness, against the spiritual host of wickedness in the heavenly places.[136]

Paul went to great length at describing evil in its many guises and was emphatic about it. The modern tendency to attribute the psychologically destructive solely to a malfunction of the human psyche is short-sighted. Chemical imbalances in the body are often much to blame and this understanding has brought great good to wellbeing. However, the insights of all major religions regarding the spirit world must not become irrelevant. Neither should it be emphasised.

The reality of sin is first mentioned in Genesis with regard to people for they are able to understand the dynamics. The creation story tells that sin is active in nature and has it function imperfectly. Its influence is depicted in the disobedience of Adam and Eve, the crown of God's creation, and its destructive consequences. A situation that would be resolved by the second Adam, Jesus Christ. Adam and Eve is the story of human reality and that of Christ. It portrays what life is about. The good and the bad.

The biblical story does not correspond with science. One obvious difference being that God created for six days while the universe is billions of years old. Herman Dooyeweerd explains.

> God's creative deeds surpass the temporal order because they are not subjected to it. But as a truth of faith God has revealed these creative deeds in the faith-aspect of his temporal order which points beyond itself to what is supra-temporal. It was God's will that the believing Jew should refer his six work days to the six divine created works and

the sabbath day to the eternal sabbath rest of God.[137]

One day is like a thousand years to God, Apostle Peter wrote; and a thousand years as one day.[138] God dwells beyond time. There is no reason why our universe could not be billions of years old, as science suggests.

A further difficulty with the creation story is that scientific data does not support of a Fall to have occurred. John Polkinghorne found the Fall to be the Christian doctrine most difficult to reconcile with scientific thought.[139] Death and disease, the influences of sin, were operative on earth long before any sign of human occupation. But Genesis records that sin entered because of Adam and Eve's disobedience and thus decay and death became a reality.

An answer to this dilemma may be found in how the first chapters of Genesis are understood and interpreted. Two creation accounts are recorded. The first in Genesis 1 till 2: 1-3. The second reads on till chapter 4.

The two accounts are partly similar in describing God's creation process and are complimentary. All is perfect, and then a Fall occurs. Is that, however, the sequence of events? Or is the focus of the first account different from that of the second? May it be that the first account reveals what God had in mind for creation; what it will end up to be like – very good! The first story tells of creatures eating every green plant for food. It reminds of Isaiah's vision in which the wolf and the lamb dwell together in peace and the lion will eat straw like an ox.[140] The story records the process by which God was to create towards the ultimate ending with a perfect humanity as its crowning glory. Sin is not considered, for it is irrelevant in the final outcome.

Once God saw that the creation was very good and at rest, God rested also, and not before. That final rest surely will be once the New Creation secured in Christ is complete.

The idea that God created for six days, is now at rest, and leaves creation to its struggles, is unsustainable. Love cannot rest until that which it loves flourishes in wellbeing. That is not presently so. The creation story presents a God at work. It should be possible for this work to exist within God's eternal rest. God is at work while at rest. Finally only the rest of God will remain. This time and space explanation has its limits. Divine reality exists beyond it.

The second account is anthropocentric. It is the story of people, of a divine promise and of Christ. It depicts the human predicament. The story refers to something basic in human experience, which thus qualifies it as a myth. Without this story humanity would be without a frame of reference for living in a world infiltrated by sin. We would be unable to understand our existence as being exposed to forces beyond universal reality, those of the spiritual realm. Nor would we know of the relational possibilities on offer. Those of companionship. In particular that of marriage; and the offer of being relational with God. This information is given against a background of how it could have been, had sin not come to spoil. The difficulties also highlight the deceptive nature of sin and its power.

God's future plan about what perfection will be like is revealed in the form of that wonderful Garden of Eden. It cannot presently be entered, for sinfulness forbids it. But a thief, who hung next to Christ on a cross, will do so. The key to the gates of Paradise is held in Christ's scarred hands. That is the promised future foretold in Genesis.

The creation account of the second story gives deep insights into the nature of people. They are formed by dust and alive through the very breath of God.

People have been created in the image of God, the best of the best, with abundant ability. It is a significant revelation contained in the narrative. However, life is described just as it is. The woman will give birth in pain and the man will toil, because of being subject to sin. The word 'fall' is not used in the story. Eve is not cursed, nor is Adam, but the ground is. The very nature of the earth has been infiltrated by sin. It is our present reality. God asks Adam and Eve whether they have eaten the forbidden fruit from the tree of the knowledge of good and evil. They obviously had. The beautiful fruit symbolises desire and its potential. It is fundamental to common human experience.[141] The other tree, the tree of life, is a promise of eternal life. Both trees became shut away in a garden. Life as it is and life as it will be, are divinely secured. Adam and Eve are locked out. On earth toil and death will be their future.

The human predicament is depicted as an obedience issue. It highlights the unique human capacity of being able to choose – between good and evil. It emphasises 'the image of God' declaration. It highlights morality and free will and shows the importance of conscience. It is this very quality that allows for a solution to the problem of sin to become possible.

For the second creation story is also that of the

second Adam. The foe to be defeated is the tempter, who is Satan. In the story he presents as a serpent, smart and wily. His power though is doomed. The serpent is cursed into the dust and the seed of the woman will bruise his head. Satan, in turn, shall bruise the heel of Christ, which occurred on the cross. God sacrificed an animal to dress Adam and Eve with garments of skin. God acknowledged the vulnerability of people. This alludes to creation existing under the covering and care of God and its promised deliverance by the sacrifice of the only Son. So much God loves the world.

In the story sin entered the world through the wrong choice made by Adam and Eve. This idea is symbolic and must be read with Christ in mind. It foretells how sin will be defeated. By a person. A human being, who Apostle Paul called the second Adam.

The knowledge of good and evil is central to the salvation narrative. As is the tree of life. The moral ability of making choices is the bedrock of humanity and also the angelic host. Angels are doing God's bidding. Satan, a fallen angel, is the destroyer. Jesus Christ, the ultimate personification of love in creation and Satan, the personification of sin, are facing off at the cross. One primary relational dynamic against another.

Jesus was victorious in this winner takes all fight.

The results of the victory were divinely planned before our present creation came about. A new heaven and earth await with the realm of angels also at rest. The two creation stories in Genesis are preludes to Good News.

Chapter 10

A relational universe

Science accepts that everything exists within cause and effect. It exists in relation; one energy dynamic always influencing another. At the level of objects it is no different. A stone presses down on the earth beneath it. The animal world functions in relation while with people it greatly determines the quality of life. Philosopher Ludwig Feuerbach (1804-1872) discussed the fact that people exist 'in relation'.

> His main argument was that the essential or typical human properties – love, reason and will – could not be understood or accounted for in terms of a single individual: they requires a minimum of two, an 'I and a Thou.[142]

It inspired Martin Buber to write his book *I and Thou* and introduce his concept of 'I and Thou and It and We'. Buber suggests that an 'I-Thou' moment, available to everyone, offers an exceptional relational experience. It may have a mystical quality something Buber would deny. His involvement with mysticism ended badly.

That Christianity is fundamentally relational is easily shown in the Great Commandment. Theology rightly places an emphasis on community as a key factor in Christian praxis. Without question, the relational is important.

The nature of spirit is love. Love is relational by nature and positively so. Sin is relational also, but destructively so. Science has sporadically addressed the relationality of our universe from a neutral perspective with the relational having no meaning beyond mere connectivity. One prominent scientist who reflected more deeply on all existing in relation was physicist David Bohm. He drew analogies between quantum mechanics and Eastern religion. Bohm's ideas had been influenced by the Dalai Lama and also Hindu mystic Krishnamurti to whom he was a friend and student. His work is an example of how spiritual understanding might be brought to bear on scientific enterprise.

Bohm is known for the de-Broglie-Bohm pilot wave theory. It proposed a way in 'reconciling the many possibilities for a quantum particle's state that are encoded in its wave function with the fact that we can only ever measure one of them.'[143] Many scientists considered multiple universes to be the answer to these quantum particle possibilities. Not everyone is convinced. In 1927, 'De Broglie maintained that every quantum particle possesses an invisible pilot wave, which runs alongside it and tells the system how to behave.'[144]

Not long after the introduction of pilot wave theory, and its accurate prediction of wave interference, the idea fell out of favour. 'It was supplanted by the view of Neils Bohr and Werner Heisenberg, who believed that the wave function encapsulates merely what we can know about reality rather than reality itself.'[145] Since 1980, pilot wave theory has attracted increasing attention again. David Bohm continued to develop the opportunities a pilot wave offered as a connection between science and the transcendent.

Bohm held that reality is a combination of two interacting orders, one implicate and the other explicate. The first is enfolded, from which the second, an unfolded order, emerges. John Horgan explains:

> Underlying the apparently chaotic realm of physical appearances – the explicate order – there is always a deeper, hidden, implicate order. Applying this concept to the quantum realm, Bohm proposed that the implicate order is a

quantum potential, a field consisting of an infinite number of fluctuating pilot waves. The overlapping of these waves generates what appear to us as particles, which constitute the explicate order.[146]

Reality as we know it is the explicate order and not only the physical side of it. In his book *Wholeness and the Implicate Order* Bohm writes:

> We are suggesting that the implicate order applies both to matter (living and non-living) and to consciousness, and that it can therefore make possible an understanding of the general relationship between these two, from which we may be able to come to some notion of a common ground of both (rather as was suggested ...... in our discussion of the relationship of inanimate matter and life).[147]

In his book, Bohm's Hindu belief is apparent. The Hindu faith never alludes to a Creator God. Generally, his theory suits the idea of spirit being central to understanding the universe. Section 8 of chapter 7, the final section of *Wholeness and the Implicate Order*, supports that suggestion.[148]

Bohm was aware of his work as just the beginning of a long journey in figuring out the correlation between the physical and that which is not. But science is uneasy about pilot waves. 'This somewhat ghostly wave that multiplies entities is one of the reasons why Bohm's theory has not commanded much support among physicists,' John Polkinghorne observes.[149] But Bohm's effort in finding a way by which a hidden influence might be incorporated into physics, must be appreciated.

The idea of the universe being relational is not a recent one. Three philosophers, Gottfried Wilhelm Leibniz, Alfred North Whitehead and currently Harold H. Oliver, have given the question attention. Their thoughts can be stated in a brief summation. A full exposition of their philosophy of the relational would be quite extensive.

Bertrand Russell considered Gottfried Leibniz (1646-1716) one of the supreme intellects of all time.[150] In mathematics, he invented the infinitesimal calculus independently from his contemporary Isaac Newton. Leibniz was convinced that a mechanical view of the universe, such as Newton's, would come up short in explaining reality adequately. He began to answer the question of existence at its most primary level. Creation originated with God, who is at One and not dualistic. Therefore Leibniz began to address the unified reality from which the universe has issued with a theory.

He felt that energy, a certain Force, was needed as a primary dynamic principle to keep the momentum of the universe alive. This energy, Leibniz decided, was to originate from primordial intelligent ideas called 'monads.' The monads, myriads of them, reside in the mind of God. They may or may not be actualised in matter. When a monad becomes actualised, when God acts out a thought, that monad will be the energy source of the resulting matter and its determining agent in how that particular matter will function.

Monads are interrelated like thoughts in a mind. Thoughts exist in relation to other thoughts and are not independent of the thought process. Leibniz decided that when monads are actualised in matter, there is no direct relation between the different kinds of matter. The relation exists only between the source of that matter's energy – the monads. These come from the Mind of God, who could potentially have created all kinds of worlds. As God will only ever create the optimum, the world we are living in is the best of all possible worlds.

Leibniz' essential contribution to metaphysics was

that matter is not lifeless, as Newton suggested. This idea was mostly ignored until in the days of Einstein, when Leibniz' theory began to find scientific relevance. Harvard professor Alfred North Whitehead (1861-1947) used his ideas for a further exposition of the relational.

For Whitehead, the relational nature of reality became the cornerstone of his metaphysics. No longer was the relational merely an aspect in explaining our universe. Instead, the relational is fully integrated with being itself. All that exists, exists as a relational expression. That relationality finds connection in divinity. The God who is, is fully related to what happens on earth. But instead of monads, Whitehead suggested 'actual entities or occasions of experience,' which are transient – very short lasting. Or, stating the same idea: the world is made up fundamentally of 'drops of experience, complex and interdependent'.[151] Interdependence holds a central place in Whitehead's thinking. It signifies a direct relatedness, which is constantly active. This interacting results in events, myriads of them. The material world is the outcome of those events which, being transient, means being in flux. In this way past becomes present, becomes future. The universe is always in progress, in the process of becoming. It is not a system of energy that is predetermined.

With Leibniz, the monads are internally related and interact in response to the 'perceptions' each detects with the other, but matter is not. For Whitehead that is insufficient. He suggests that 'actual entities or occasions of experience' also called 'drops of experience,' are both internally *and* externally related. When these *internally* related 'drops of experience' interact, it creates 'an event.' That event prehends other events, as Whitehead describes it. This 'prehension' is relationally *external*. 'Each individual event prehends all the other events of the world that it knows.'[152] And thus, all is relational. Leibniz wrote about matter being 'ensouled', but itself not directly inter-relational. In Whitehead's theory matter is relational in itself – everything is.

The word *thing*, as in everything, is what contemporary theologian Harold Oliver takes issue with. Thing refers to a subject/object relationship. Both Leibniz' and Whitehead's theories take that approach. Oliver proposes a system in which relation exists prior our subject/object reality. Rather than being an outcome of two things interacting, it should be that those two things exist as the expression of relation. The relational is primary and prior to any thing.

Oliver finds the current interest by scientist in relationality encouraging and mentions David Bohm as an example. Oliver suggests the building block of the universe to be called *reals*. It is representative of the actual relational itself. As such it is transpolar – beyond polarity.[153] Subject/object are preceded by relation. The relational, as we known it in everyday life, is a derivative of a prior dynamic. Oliver wrote his idea down in the three-letter description: **aRb**.

The **R** in this description is Relation, with **a** and **b** its functional dependencies. These dependencies come about because of **R**.[154] Relations themselves do not move and are at rest. They are, what Oliver calls, *pure activity*. Like being, which just *is*. When this *pure activity* is stimulated, it can issue *relatio* as a concrete action and that dynamic manifests as substance. This action in response to stimulus can be 'acting on' and 'acted upon', for the relational reciprocates. The two actions are represented by **a** and **b** in **aRb.** They are inseparable from **R** and are its derivatives. Neither **aR** nor **Rb** is possible. Ontologically, it is the 'acting', the *relatio*, that is real and it results in our reality.[155] For Oliver our universe is not merely relational, it issues from relation. Leibniz' and Whitehead's concepts involving internal and external relating become superseded. Oliver considers the relational primary. Only relation is real, everything is derived from it and in a manner that never loses its structure and purpose. Even time and space are derivatives.[156]

In making relation primary to everything Oliver holds that even our idea of God is merely a name for the 'Totality of Relations' which is 'the composite of all relations.'[157] This Totality originates from the truly real. God is not first, relation is. This idea, and his Christology in which Christ is not God, is unacceptable. Oliver's philosophical creativity though is interesting. It may not be surprising that of late he has become deeply interested in Buddhism, which does not recognise a deity.

Martin Buber, in *I and Thou*, tells how a subject/object relation might be overcome experientially in people. 'I/Thou' is a relational event that can happen to people occasionally when suddenly the realm of 'between' is experienced. The 'I' exists in our natural reality. 'Thou' is beyond space and time. It is the eternal Thou. The 'I-Thou' moment occurs when a meeting between two people, or a person and a tree for that matter, turns into a revelation of interconnectedness that reaches beyond the usual subject/object perception. Buber explains that

Thou finds its context 'in the Centre, where the extended lines of relations meet – in the eternal Thou.'[158]

In an 'I-Thou' experience the perceptive restrictions of personhood are overcome and the relational in its all-encompassing nature is unveiled. This experience cannot be personally activated but may somehow be spiritually bestowed.

The primary dynamics of relation are found in spirit. By spirit all exists and its relational influence on creation is twofold. Spirit is the agent of love and all that entails. But sin is allowed to operate also. Two opposite powers. The mystery of that remains with God. The experience of it belongs to nature and to people. People are able to express the relational in the highest form creation allows.

That expression is always a mixture between love and sin. The spirit in every person is so imprinted. Having a spirit means being relational for spirit and relation are indivisible. What is called spirituality is in essence the same as relationality. Every person has a spirituality expressed in accordance with the inner nature of that person. This nature is relationally defined. Spirituality is a relational construct that may include religious belief. The belief itself does not determine the quality of a person's spirituality. The attitude and actions of a person do. A believer may seem highly spiritual while misbehaving, being egotistic and the like. The actual spirituality of that person will be of lesser quality than that of a non-believer who lives compassionately and with integrity. Motivation and intent are the markers of value. That applies to any race, colour or creed. It is what God will appraise.

The relational is also a determinant in wellbeing. Everyone has the influences of love and sin within their person. They affect not just people but groups, societies and countries. At a personal level much is decided by the relational imprint obtained in growing up. Relation is central to everything. When something goes seriously wrong, there usually are destructive relational influences involved. When life's circumstances are psycho-spiritually healthy and optimistic, the empowerment of love is having the upper hand. It is not here to explain this in detail.[a]

The central place of relationality in human affairs must not be underestimated. The very fabric of creation exists in relation. We are living in a relational universe.

---

[a] See, *The Primacy of Love,* Michael J Spyker, AgapeDeum

Chapter 11

Science and Spirit

In creation there is no disharmony between nature and spirit. They present as one homogenous whole. Both nature and spirit are merely words that signify realities. These words have come about because people have perceived the actuality of what they represent. That perception was not intellectual, but came from within the depth of their being. Every religion and cult testified to it and still does. If science restricted itself to the business of investigating nature, and recognised spirit in its rightful place, there would not be a problem. But certain scientists influencing society have overstepped that mark. They argue that nature and spirit cannot be at one for the idea of spirit is an illusion. Spirit is no more than an expression of nature itself. Molecular biology is at the forefront of this. Francis Crick, known regarding the discovery of DNA, made this explicit in his book *Of Molecules and Men* (1966). He wrote that, 'the ultimate aim of the modern movement in biology is in fact to explain *all* biology in terms of physics and chemistry.'[159] The aim is fine, but its success when ignoring the reality of spirit will be partial.

The Enlightenment liberated the Western mind years ago. Today, technology is liberating consciousness, the human ego. A new sense of security is being felt. Old taboos are overthrown one by one. Technology offers certainty and control. Human destiny rests in the hands of scientists and technicians.

The ability of people and access to information and entertainment has massively increased. There is no need to ever be bored or attend to the inner self with the complex questions that can bring. There are of course insecurities in life for nature is unpredictable. So are people, with hassles everywhere. Still, I may live in the confidence of being me. The kind I am daily deciding on.

This shift in lived experience has been noted in psycho-analyses. This science was first introduced by Sigmund Freud (1856-1939) and further developed by others. Psychoanalysis teaches that the psyche functions by the interaction of three entities. Jones and Butman explain:

> The *id* is the repository of our most primitive sexual and aggressive drive and urges. ….The *ego* is reality-based and develops as an individual interacts with the external world. …. Finally, the

> *superego* places restrictive demands on both the id and the ego ….. It is generally understood to be … the repository of the moralistic standards one has absorbed from "significant others".[160]

Jones and Butman add that the id processes are largely unconscious, illogical and relentlessly driving towards personal gratification. The ego is predominantly conscious and mediates with the primitive urges of the id. The superego is assumed to be partially conscious and is seen as a kind of 'conscience'.

A power shift between the id, ego and superego, defines the modern person from the past. The influence of the superego is diminished. The id, which includes the pleasure principle, has been led off the leash. The ego has become the main focus of what individuality means with few restrictions. This change has occurred because of people having faith, not in religion, but in the secular mind. Psychiatrist Robert Coles, professor at Harvard, observes that,

> …. the faith we have in science; the rise of the social sciences, and their sense of entitlement with respect to the credibility extended them, the expectations entertained of them – all of that being evidence of the faith we have in ourselves, in our

> ability to know ourselves, gain control of things (within and outside ourselves) through such knowledge. …. It is a faith in the capacity of the human brain (the organ that has investigated successfully all other organs) to explore itself, understand itself fully, gain operating (clinical) control over its vulnerabilities, aberrancies.[161]

The effects on consciousness are noticeable. David Brooks writes that the shape of human life was altered by a shift to the Big Me – a new mindset based on the romantic idea that each of us has a Golden Figure in the core of ourselves. He quotes philosopher Charles Taylor who writes,

> There is a certain way of being that is my way. I am called to live my life in this way and not in imitation of anyone else's …… If I'm not, I miss the point of my life. I miss what being human for me is.[162]

In his book *The Road to Character* Brooks uses an idea by Rabbi Joseph Soloveitchik to frame his presentation. 'The Rabbi noted that there are two accounts of creation in Genesis and argued that these represent the two opposing sides of our nature, which

Science and Spirit 137

he called Adam 1 and Adam 2.'[163] Brooks interprets that as Adam 1 being the career oriented, ambitious side of our nature. Adam 2 is the internal Adam. *The Road to Character* was a bestseller, which shows that not everyone is content with the overly materialistic nature of our society.

In today's culture the prophets of scientific sufficiency find a ready hearing. People find solace in the idea of not having to worry much about the spiritual. Until faced with a death situation when insecurities can arise. Those soon are placated by the dynamics of a life that goes on. The sufficiency of science appeals to people who like their intellectual capacities set free from the God question. Scientists Stephen Hawking (recently deceased) and Richard Dawkins are two major proponents of the view that God now is irrelevant. They exert considerable influence.

Dawkins' influential book, *The God Delusion* (2006) offers a scintillating critique on the history of the Christian religion and the grievous harm it has inflicted through the ages. He promotes the sciences and shows the advantages of non-belief to society and individuals. Not surprisingly, the book found many readers. Dawkins designed a spectrum of seven probabilities about the existence of God. It ranges from 'I know there is a God,' down to, 'I know there isn't one.' He rates himself at number six: 'I think God's existence is very improbable.' It makes him not quite the atheist we are made to believe, but an agnostic.[164] But does it? As a scientist dealing with the rigour of facts and probabilities, and in the knowledge that the existence of God cannot be proven either way, Dawkins has to remain true to the perspective that perhaps there might be a God, however small that probability. Dawkins' mind is philosophically arrested by the dictates of logic regarding God. His heart though has no such restrictions, which his writings confirm.

Hawking and Dawkins are a manifestation of an old phenomenon. Voltaire ridiculed the church long before and Nietzsche declared God dead in emphatic fashion. The difference is that these days such declarations are not philosophically presented as correct but scientifically. At least, it is stated thus. The truth is different. God cannot be declared dead scientifically for science does not investigate God. Rather, it investigates God's handiwork. While any assertion about faith, by someone who does not firmly believe in God at heart, will by necessity be philosophical.

If Dawkins is to be corrected, it must be on the

wrong impression he gives about scientific certainty. The authority of the sciences can and must be questioned. Rupert Sheldrake in his book *The Science Delusion* does exactly that. He highlights ten core beliefs that science takes for granted but which need further scrutiny.

1. Everything is essentially mechanical.
2. All matter is unconscious.
3. The total amount of matter and energy is always the same.
4. The laws of nature are fixed.
5. Nature is purposeless.
6. All biological inheritance is material.
7. Minds are inside heads and are nothing but activities of brains.
8. Memories are stored as material traces in brains and are wiped out at death.
9. Unexplained phenomenon like telepathy are illusionary.
10. Mechanistic medicine is the only kind that really works.[165]

Sheldrake insists that the scientific mind must be liberated from these restrictions for science to move onward.

In nature the findings of science and the reality of spirit are in harmony, they are at one. Any perceived dualism is grounded in human experience and the will. If it is willed that unity is detected, then that is possible. It means giving credibility to both scientific and revelatory insight. Diminish or ignore one of those and the actual harmony between science and spirit cannot be found. The sciences are incredible, while religion has a wealth of knowledge and valuable wisdom. Reason cannot singularly fathom the mysteries of the universe. It needs the assistance of the spiritual and vice versa.

What prevents a perceived unity between science and spirit is not scientific nor religious. It is philosophical. How is my worldview being challenged? The worldview I seek to believe in and that my ego is cultivating. This problem fires the science and religion question to a large degree. There is more than knowledge involved and includes the subjectivity of the knower. Science is a major cultural influence. It needs theological attention. Of a kind that is well informed, creative and non-defensive. Entrenched positions are unhelpful.

The Christian scientist need theology's help. As does the everyday Christian. Effort must be made to engage across the divide. Dogmatic orthodoxy may hinder this process and where necessary should be challenged. The languages of science and religion are different. But connectivity will improve when seeking ideas that might be relevant both to Scripture and physics. Perhaps theology should aim to speak into the 10 point plan that Rupert Sheldrake has suggested for the sciences. It might test itself whether in this challenge it can be effective. Not just theology speaking to theology, but communicating meaningfully with the scientific community. Involve the Christian scientists in finding answers. Undoubtedly, that will be appreciated by many.

The main problem the sciences present to religion is not only one of compatibility. The challenge exists on a much broader scale and is one of influence. In society science has stolen religion's thunder. Christianity has become irrelevant because of scientific certainty. And yet, it has an ace up its sleeve. Christian belief is essentially a matter of the heart. It offers heart supported by reason. Modern society is not in need of reason, there is enough of it about. But it needs guidance with answers for the heart. Significant heart questions tend to be avoided by people these days. The Church might help change that.

Its call in modern society mostly falls on deaf ears. The Church has lost respect and finds it difficult to frame a message that resonates. It likes to speak from the heart but comes across as ritualistic, historical and legalistic. The call does not have the right 'feel' about it to attract society's attention. Communicating matters of spirit in a self-sufficient world is difficult. It needs deep thought and a clear understanding of the task ahead. The present renaissance of spiritual theology is a fine step forwards. That branch of theology is still underrepresented in Christian study. It is also underestimated.

Society must be offered the heart of Christ. Not thereby avoiding dogmatics, but making sure it is of a kind that knows Christ's heart from lived experience. Thomas Aquinas is an example, which is not always realised. His massive intellect was submissive to the affective nature of his spirit towards God. His prayer experiences bordered on mysticism. It depends on your definition. Today's message to society must major on God's love and embrace humanity without reserve. That is not how the Gospel is generally being perceived.

People still associate the Gospel more with sin than with love. Church practices have historically left the wrong impression. 'I was born an original sinner,' Annie

Lennox sings scathingly. Add to that the increasingly reported misbehaviour of ministers and clergy and the picture becomes unattractive.

A sin focussed message, even when tempered by the declared love of Christ, is unsuited to modern times. It is questionable whether a theology that places major emphasis on sin was ever the correct representation of the Good News. The Gospel is not an accusation but an invitation.

God *is* Spirit. God *is* Love. God's creation exists in universal spirit and its nature is love. Sin has been an intruder into this from the start of creation. But sin has legitimacy before God. 'Sin is necessary,' Jesus told Julian of Norwich. To the question as to why, she would not get an answer. The Son of God was sacrificed because of this necessity of sin. Obviously, the reality of sin thus features prominently in the Gospel story. In response to the plight of humanity, the Church has declared its sinful state and what that brings. Theologically, it is correct but for one point. Whenever considering sin, the love of God should be mentioned as more powerful and important. The Gospel story is one of love, with sin not deserving the attention often given to it. Many clergy and believers would agree with that.

When I became a Christian years ago at a revival

meeting extolling the Lord, the Pastor prayed over me thanking God for accepting this sinner. I distinctly remember thinking, 'what's that about?' Me a sinner? I thought to have lived a fairly decent life thus far. Of course, later, I well understood. I also understood that the concept of sinner can be a real communicative stumbling block. More so today even than during my twenties. I would much prefer using the idea of 'coming up short.' As Apostle Paul so passionately declared: 'All have sinned and come short of the glory of God.'[166] The Law was given to show humanity up as sinners, Paul explained. Everyone is a sinner.

That simply is a fact – no arguing about it. But is it also an accusation? I'm not so sure. Even if it might be interpreted as such, what are the benefits in blaming people for something they are inherently born with and that has them make mistakes? Yes, it may fosters guilt, but is that what a loving Heavenly Father seeks? Not according to the parable of the Lost Son.[167] I prefer to believe that Paul used the sin-fact not to highlight an accusation but as an encouragement towards accepting the great gift of Jesus Christ: the saving grace on offer because of God's love for the world. Furthermore, in an outreach that properly understands the nature of love, power games that can infiltrate as a meta-message in the presentation of Christ will become impossible.

Not just those who don't know Christ, but believers themselves may suffer if the place of sin in the Gospel is inappropriately presented. They may sense that the label of 'sinner' cannot be escaped from, even if one is saved by grace. They may be unable to shake of a feeling of unworthiness, now so clearly stated. Perhaps because of a troubled self-perception that began at childhood. So, the gift of grace that is meant to liberate from such ensnarement is actually enforcing the problem.

At the other side of the spectrum, people who have correctly accepted that they are now covered by grace may become filled with pride at this superior position in life. The pitfalls of allowing sin to overshadow the greater power of love are many and ever present. It needs careful handling.

Sin is necessary, but pales into insignificance before the eternal love of God. Jesus never focused on sin, but on the goodness of his Father. He highlighted how a good life can be lived, showed up hypocrisy mainly amongst the ruling classes, had compassion on all and the poor in particular. As the Christ, he came to cover over the ravages of sin by love and thus destroy its power. The Good News is not one of condemnation but of divine understanding. The Gospel of Jesus Christ, unhindered by religious trappings and specialised teaching, is a message of love meant to speak directly to the heart. It has the ability to penetrate the modern psyche. It is not primarily concerned with the dos and don'ts of life. Those are intrinsically known anyway. The Gospel seeks to bring a person into relationship with God, in spite of everything that person might be – good and less admirable. Once the love of God has entered someone's life, spiritual awareness will sharpen and life will be strengthened. Scripture will become revealed and meaningful.

God's love is magnificent. The full nature of it cannot be understood. Occasionally, the mystics have had a taste and found no words by which to describe the depth of their experiences. This love holds creation together with the sure promise of an incredible eternal future. As a seed cannot imagine the beauty of its flower to come, so we cannot the New Creation. That is the Gospel story.

If science has stolen Christianity's thunder, it should be possible to regain some ground. There is no need to be on the defensive. Christians may know that science itself is not a threat. It is wonderful and God's gift to humanity. Christians will do well in having some knowledge of scientific achievements. Christian scientists may be affirmed in knowing that there is no actual demarcation between science and spirit, but only a perceived one. Theology must be creative and increase its relevancy. Hurdles and opportunities should be clearly noted and addressed in the confidence of God's spirit. Bridges must be built. Harmless as a dove but with penetrating insight. Faith taking courage.

> In a time of drastic change one can be too preoccupied with what is ending or too obsessed with what seems to be beginning. In either case one loses touch with the present and with its obscure but dynamic possibilities. What really matters is openness, readiness, attention, courage to face risks. You do not need to know precisely what is happening, or exactly where it is all going. What you need is to recognize the possibilities and challenges offered by the present moment, and to embrace them with courage, faith and hope. In such an event, courage is the authentic form taken by love. (Thomas Merton)[168]

Science and Spirit has dealt with possibilities and problems across a broad spectrum. As a short treatise it is limited to superficial detail. In the scientific aspects I could not have offered better in any case. *Science and Spirit* is a thought exercise. Meant as a step in the right direction, no more. My aim has been to inform, question, and stimulate. Whatever the outcome: may the heart find its rightful place in the thoughts of the head. May spirit and science exist in harmony.

Thank you for reading this book

I hope, you found it interesting

For more, please read on.

Many blessings, Michael

Books by Michael J Spyker
Available at agapedeum.com

Trilogy

Meeting Emma

A journey of discovery in which Emma becomes familiar with the many idea of Christian Spirituality through the ages. It helps her towards the person she would like to be. This book has assisted many in coming to love the vast wealth of the Christians spiritual tradition.

The Primacy of Love

Jake hears about his father's ideas on God's Love from Baz while travelling the Simpson Desert. Their talks include the significance of eternal and universal love, and the relational. The story has been called a significant theological feat.

The Language of Love

Emma and Jake fall in love. JH introduces them to the real meaning of Eros well beyond merely sex. They learn about being a Friend of Jesus and the language of love. Emma and Jake set off camping in the outback in search of JH. They work out what it means to live intimately together.

Novels

Julian's Windows

A musician and a teacher of children with intellectual ability fall in love. He lost his wife. She questions her vocation as a religious sister. Country life in Victoria restores his soul. A holiday in Australia from Liverpool decides her future. The ideas of Lady Julian of Norwich are an integral part of this love story in a most natural way. Great fun and informative.

Shalomat

Jacq and Ahmed, 16 years old, are on the run through Australia on a quest with mystical dimensions. It draws them together. All seems lost but isn't quite. Young people and adults enjoy this adventure. It is partly a comment on the one-sidedness of modern society and uses ideas of spirituality and philosophy. Will there be a sequel, an appreciative reader asked?

Treatise

Science and Spirit

Science exists by the creativity of God. But where to find God within physics? Where in society, in which God has become irrelevant? An informed answer best includes knowledge of history, science, philosophy, theology and religion. Plus ideas about a way forward. A read of significance to enjoy.

Christian Living

Drawings and Reflections

52 short reflections and 16 drawings that lift the spirit. A brief story that sows an idea. A picture to enjoy. It is not so easy to stay focused in a busy world. A little help always comes in handy. There is nothing religious about this book apart from keeping Jesus in mind and living vibrantly.

Chapter 1

[1] Spengler, O 1962, *The Decline of the West*, Vintage Books, New York p. 322

[2] Westfall, RS 1993, *The Life of Isaac Newton*, Cambridge University Press, Cambridge p. 124

[3] Popper, K 1976, *Unended Quest*, Fontana/Collins, Glasgow
p. 41

[4] Dooyeweerd, H 1960, *In the Twilight of Western Thought*, Craig Press, New Jersey p. 6

[5] Ibid, p. 7

[6] Gribbin, J 2012, *Erwin Schrödinger and the Quantum Revolution*, Bantam Press, London p. 289

Chapter 2

[7] Blatner, D 2012, *Spectrums, our mind-boggling universe from infinitesimal to infinity*, Bloomsbury p.37

[8] Ibid, p. 48

[9] Ibid, p. 48

[10] Colossians 1:15-17

[11] John 4:24

[12] Smolin, L 2008, *The Trouble with Physics*, Penguin Books, p3

[13] Kumar, M 2014, *Quantum*, Icon Books, London p.p. 140-141

[14] Smolin, L 2008, *The Trouble with Physics*, Penguin Books, p6

[15] Kumar, M 2014, *Quantum*, Icon Books, London p. 329

[16] Blatner, D 2012, *Spectrums, our mind-boggling universe from infinitesimal to infinity*, Bloomsbury p.123

[17] Ibid

[18] Smolin, L 2008, *The Trouble with Physics*, Penguin Books, p. 316

[19] Dooyeweerd, H 1960, *In the Twilight of Western Thought*, Craig Press, New Jersey p. 163

[20] Armstrong, K 2007, *The Great Transformation, the World in the Time of Buddha, Socrates, Confucius and*

Jeremiah, Atlantic Books UK p.330

21 Descartes, R 1968, *René Descartes – Discourse on Method and The Meditations*, Penguin Books p.p. 176-177

22 Meister Eckhart 1994, *Meister Eckhart – Selected Writings*, Penguin Books, p. 20

23 1John 4:8

Chapter 3

24 Bartusiak, M 2015, *Black Hole,* Yale University Press, London p.112

25 Blatner, D 2012, *Spectrums, our mind-boggling universe from infinitesimal to infinity*, Bloomsbury p.45

26 Ibid

27 Cousins, E 1992, *Christ of the 21st Century*, Element, Rockport, Massachusetts p. 5

28 Ibid, p. 6

29 Sharma, A 1993, *Our Religions*, HarperSanFrancisco p. 21

30 Griffiths, B 1992, *A New Vision of Reality,* Fount – HarperCollins, London p. 65

31 Griffith, T (ed.) 2000, *Upanishads,* Wordsworth Editions Limited, Hertfordshire p. xiii – xiv

32 Griffiths, B 1994, *Universal Wisdom- a Journey Through The Sacred Wisdom Of The World,* Fount, London p. 175

33 Cousins, LS - Hinnells, JR (Ed.) 1984, *A Handbook of Living Religions*, Penguin Books p.305 -306

34 Abe, M - Sharma, A (Ed.) 1993, *Our Religions*, HarperSanFrancisco p. 76

35 Nasr, SH - Sharma, A (Ed.) 1993, *Our Religions*, HarperSanFrancisco p.462

36 Bloemendal, M – Fennema J & Paul I (eds) 1990, *Science and Religion*, Kluwer Academic Publishers, Dordrecht p. 56

37 Copleston. FC 1972, *A History of Medieval Philosophy*, Methuen & Co Ltd, London p. 278.

38 Ibid, p. 287

[39] Polkinghorne, J – Fennema J & Paul I (eds) 1990, *Science and Religion*, Kluwer Academic Publishers, Dordrecht p. 88

Chapter 4
[40] Cousins, EH 1992, *Christ of the 21st Century*, Element, Brisbane Queensland p.p. 6-7
[41] Stendahl, K 1976, *Paul Among Jews and Gentiles*, Fortress Press, Philadelphia p. 85
[42] Norrestrander, T 1999, *The User Illusion,* Penguin Books, p. 320
[43] Horgan, J 1996, *The End of Science*, Little, brown and Company, London p.181
[44] Nagel, T, Honderich, T (ed) 2015, *Philosophers of our Times*, Oxford University Press, UK p. 19
[45] Polkinghorne, J 1991, *Reason and Reality*, SPCK, London p. 181
[46] Searle, JR, Honderich, T (ed) 2015, *Philosophers of our Times*, Oxford University Press, UK p. 217
[47] Norrestrander, T 1999, *The User Illusion,* Penguin Books, p. 326
[48] Sheldrake, R 2013, *The Science Delusion*, Coronet, London, p. 228
[49] Ibid, p. 227
[50] Ibid, p. 214
[51] Ibid, p. 222
[52] Pannenburg, W 1991, *Systematic Theology* vol. 1, Eerdmans, Grand Rapids p.p. 382ff
[53] Chalmers, R, Honderich, T (ed) 2015, *Philosophers of our Times*, Oxford University Press, UK p. 351

Chapter 5
[54] Dooyeweerd, H 1960, *In the Twilight of Western Thought*, The Craig Press, New Jersey p.179
[55] Delitzsch, F 1867, *A system of Biblical Psychology*, T&T Clark, Edinburgh p. 264
[56] Genesis 2:7

[57] Genesis 6:17
[58] Information taken from Beck, JR & Demarest, B 2005, *The Human Person in Theology and Psychology*, Kregel, Grand Rapids, p.p. 131-132
[59] Delitzsch, F 1867, *A system of Biblical Psychology*, T&T Clark, Edinburgh p. 94
[60] Ecclesiastes 12:7
[61] Delitzsch, F 1867, *A system of Biblical Psychology*, T&T Clark, Edinburgh p. 97
[62] Ibid, p. 469
[63] Romans 8:19-20
[64] Fullyawaremind.com/death
[65] Ibid
[66] Rahner, K 1964, *The Eternal Year*, Helicon, Baltimore p. 92
[67] Brueggemann, W 1982, *Living Toward a Vision, Biblical Reflections on Shalom*, United Church Press, New York, p. 39
[68] John 20:22
[69] John 3:5-6
[70] John 16:7
[71] Acts 19:2
[72] Pannenburg, W 1991, *Systematic Theology* vol. 1, Eerdmans, Grand Rapids p.p. 382ff
[73] John 15:19; John 17:14
[74] John 17:14
[75] Brunner, E 1939, *Man in Revolt*, Lutterworth Press, London
p. 102
[76] Gen. 1:26, *The Message*
[77] Rahner, K 1964, *The Eternal Year*, Helicon, Baltimore p. 94

Chapter 6
[78] Brooks, D 2015, *The Road to Character*, Penguin Books p. 237

[79] Gribbin, J 2012, *Erwin Schrödinger and the Quantum Revolution*, Bantam Press, London p. 44

[80] Greene, B 2000, *The Elegant Universe*, Vintage, London p. 5

[81] Smolin, L 2006, *The Trouble With Physics*, Penguin Books, London p. xvi

[82] Ibid p.p. 4-5

[83] The Quantum World, New Scientist Collection, vol. 3 / issue 3 2016, Reed Business Information, England p. 13

[84] The idea of 'at the same time' is questionable.

[85] The Quantum World, New Scientist Collection, vol. 3 / issue 3 2016, Reed Business Information, England p. 16

[86] Ibid, p. 16

[87] Smolin, L 2006, *The Trouble With Physics*, Penguin Books, London p. 6

[88] The Quantum World, New Scientist Collection, vol. 3 / issue 3 2016, Reed Business Information, England p. 12

[89] Horgan, J 1996, *The End of Science*, Little, Brown and Company, London p. 57

[90] Ibid, p. 58

[91] Kenny, A 2010, *A New History of Western Philosophy*, Clarendon Press, Oxford p.995

[92] James, W 1982, *The Varieties of Religious Experience*, Penguin Books, New York p. 436

[93] Russell, B 1946, *History of Western Philosophy*, Routledge, London p. 743

[94] Ibid, p. 744

[95] Butler-Bowdon, T 2013, *50 Philosophy Classics*, Nicholas Brealey Publishing, London p. 259

[96] Romans 1:19-20

[97] E.g. J. Wentzel van Huyssteen, Nancey Murphy, Philip Clayton, Mark Worthing

[98] Torrance, TF – Fennema, J & Paul, I (eds.) 1990, *Science and Religion*, Kluwer Academic Publishers, Dordrecht p. 40

[99] E.g. Torrance, TF 1998, *Christian Theology and Scientific Culture*, Wipf and Stock Publishers, Eugene –Origan p. 9

Chapter 7

[100] Moore, S 2007, *The Contagion of Jesus*, Darton Longman Todd, London (back cover)

[101] Romans 4:17

[102] Hawking, S 1988, *A Brief History of Time*, Bantam Books NY chap. 3

[103] Kenny, A 2012, *A New History of Western Philosophy*, Clarendon Press, Oxford p. 578

[104] St Augustine 1961, *Confessions*, Penguin Books, London p. 264

[105] Ibid, p. 266

[106] Bergson, H 2004, *Matter and Memory*, Dover Publications, Mineola, New York p. 75

[107] Norretranders, T 1991, *The User Illusion*, Penguin Books, New York p. 125

[108] Smolin, L 2006, *The Trouble With Physics*, Penguin Books, London p. 321

[109] New Scientist, The Collection Issue Two 2014, *The Unknown Universe,* Chatswood, NSW p. 95

[110] New Scientist, The Collection Vol 3/ Issue 3 2016, *The Quantum World,* Chatwoods, NSW p. 114

[111] Scientific American, Volume 314, no 5, May 2016 p. 45

[112] www.technologyreview.com/s/609451/ibm-raises-the-bar-with-a-50-qubit-quantum-computer/

[113] 2 Corinthians 12:2

[114] Revelations 21:21

[115] 2 Peter 3:8

Chapter 8

[116] The Classics of Western Spirituality 1979, *Richard of St. Victor*, Paulist Press, New York p.370

[117] St John of The Cross, 1991. *The Collected Works of St John of the Cross,* ICS Publications, Washington p. 54

[118] Grant, P 1985, *A Dazzling Darkness – an Anthology of Western Mysticism*, Fount, London p. 297

[119] John 1:2; Colossians 1:15-17

[120] Philippians 2:7
[121] Butler-Bowdon, T 2007, *50 Psychology Classics*, Nicholas Brealey Publishing, London p. 143
[122] Ibid, p.p. 145-146
[123] Moore, S 1980, *The Fire and the Rose*, Darton, Longman & Todd, London p. 39
[124] Badiou, A 2012, *In Praise of Love*, Serpent's Tail, London p.p. 21-22
[125] Ibid, p. 25
[126] Brooks, D 2015, *The Road to Character*, Penguin Books p. 170
[127] Lewis, CS 2017, *The Four Loves*, Harper Collins
[128] Matthew 5:44

Chapter 9
[129] Romans 8:20-21
[130] Julian of Norwich, *Showings*, 1978, Paulist Press, New York p. 148
[131] Ibid
[132] 2 Peter 3: 7&10
[133] Bartusiak, M 2015, *Black Hole*, Yale University Press, New Haven p. 137
[134] Griffiths, B 1989, *A New Vision of Reality*, Fount. London p. 48ff
[135] Isaiah 14:12
[136] Ephesians 6:11-12
[137] Dooyeweerd, H 1960, *In the Twilight of Western Thought*, The Craig Press, New Jersey p.p. 150-151
[138] 2 Peter 3:8
[139] Polkinghorne, J 1991, *Reason and Reality*, SPCK, London p. 99
[140] Isaiah 11:6-7

Chapter 10
[141] Consider Buddhism
[142] Kamenka, E 1970, *The Philosophy of Ludwig Feuerbach*, Routledge & Kegan Paul, London p. 121

[143] Ibid, p. 60
[144] Ibid, p. 60
[145] Ibid, p. 60
[146] Horgan, J 1996, *The End of Science*, Little, Brown and Company, London p. 87
[147] Bohm, D 1980, *Wholeness and the Implicate Order*, Routledge Classics, London p. 249
[148] Ibid, p.p. 262ff
[149] Polkinghorne, J 1991, *Reason and Reality, the Relationship between Science and Theology,* SPCK, London p. 89
[150] Russell, B 1946, *History of Western Philosophy*, Routledge Classics, London p. 531
[151] Whitehead, AN 1929, *Process and Reality: an essay in cosmology*, Free Press p. 23
[152] Pannenberg, W 1990, *Metaphysics and the Idea of God*, W.B. Eerdmans, Grand Rapids p. 118
[153] Oliver, HH 1981, *A Relational Metaphysic*, Martinus Nijhoff Publishers, The Hague p. 95
[154] Ibid. p.p. 154-155
[155] Ibid. p.156
[156] Ibid. p.156
[157] Ibid, p.p. 166&170
[158] Buber, M 1958, *I and Thou,* Charles Scribner's Sons, NY p. 100

Chapter 11
[159] Sheldrake, R 2013, *The Science Delusion*, Coronet, London p. 45
[160] Jones, SL & Butman, RE 1991, *Modern Psycho-Therapies*, Inter Varsity Press, Illinois p. 68
[161] Coles, R 1999. *The secular mind*, Princeton University Press, New Jersey p. 116
[162] Brooks, D 2015, *The Road to Character*, Penguin p. 149
[163] Ibid, p. ix
[164] Kenny, A – (ed) Honderich T 2015, *Philosophers of our times*, Oxford University Press p. 266

[165] Sheldrake, R 2013, *The Science Delusion*, Coronet, London p.p. 7-8
[166] Romans 3:23
[167] Luke 15:11-32
[168] Merton, T 1965, *Conjectures of a Guilty Bystander*, Doubleday, NY p. 208

www.ingramcontent.com/pod-product-compliance
Lightning Source LLC
Chambersburg PA
CBHW072006290426
44109CB00018B/2146